# Breaking Through

## A Half-Century of the George R. Brown School of Engineering and Computing at Rice University

Written by Joe Nick Patoski

With contributions by Jada Crawford, Emily Person, and Veronica Tremblay

Rice University School of Engineering and Computing

“

It demonstrates all the attitudes an object can take, You can lean it, lay it down, or stand it up...

The *45°* element is dynamic, leaning, with the potential to continue to move.

**Michael Heizer, 1985**

on the *45°, 90°, 180°* granite monoliths in the Engineering Quadrangle

Commissioned by Alice Pratt Brown in honor of George R. Brown

Contents

# ↗ Foreword

**Ann and John Doerr**
Trustee Emeriti, Alumni, and Benefactors

Engineering has been at the heart of Rice University for more than a century. This year, we proudly celebrate the fiftieth anniversary of the formation of the George R. Brown School of Engineering and Computing, honoring five decades of groundbreaking achievements. As alumni of this remarkable institution, we feel immense gratitude for the opportunity to reflect on the school's journey—a journey that has shaped not only our lives but also the future of engineering, computing, and leadership.

Our time at Rice laid the foundation for everything that followed. The rigorous education, mentorship from brilliant faculty, and camaraderie with fellow students instilled in us a deep appreciation for innovation, integrity, and service. These values have guided our personal and professional paths, and they remain at the core of the school's mission today.

While we fondly reflect on our time at the Abercrombie Engineering Laboratory, we are delighted to see how much the school has evolved since we were undergraduates—expanding its reach, deepening its impact, and preparing students to lead in a rapidly changing world. One example is the Rice Center for Engineering Leadership (RCEL), which has empowered students to lead with empathy, clarity, and purpose. Many of its graduates have gone on to head global projects, including humanitarian efforts through Engineers Without Borders. Similarly, Rice's OpenStax initiative to support education beyond its walls reflects its enduring commitment to increasing access to the highest quality of education. These programs reflect our conviction that technical excellence must be paired with strong leadership to truly make a difference.

Over the decades, remarkable individuals have helped shape the George R. Brown School of Engineering and Computing, including Sidney Burrus, a pioneer of digital signal processing, and Ken Kennedy, the founder of Rice's computer science program and a global leader in high-performance computing. Sidney and Ken remained trusted advisors long after we graduated. We had the opportunity to endow chairs in their names to honor their legacy of teaching, mentorship, and innovation, exemplifying the spirit of the school, where excellence is not just pursued but generously shared.

We believe deeply in the importance of expanding capacity in computing, a field that is transforming every facet of society—from healthcare and energy to education and ethics. We were thrilled when, in 2024, "computing" was added to the name of what is now the George R. Brown School of Engineering and Computing.

This commemorative book—designed as a coffee-table history volume—is more than a celebration of the past. It is an inspirational reflection of the legacy and community that define the George R. Brown School of Engineering and Computing. Its pages convey the school's spirit and trace its evolution. It is a reflective yet not exhaustive example of the people and partnerships that have shaped the school over the past fifty years, inviting readers to reflect on the values and vision that will guide its next fifty.

The school's greatest strength has always been its people. Faculty who inspire. Students who challenge the status quo. Alumni who lead with purpose. And partners who believe in the school's mission. Together, this community has built a legacy of impact that reaches far beyond campus.

To the next generation of students and scholars: You are inheriting a legacy of excellence, and you are the stewards of its future. The challenges and opportunities ahead—climate, healthcare, artificial intelligence, sustainable infrastructure, and global equity—demand bold thinking and compassionate leadership. We are confident that the George R. Brown School of Engineering and Computing will continue to rise to meet them.

To our fellow alumni, faculty, and friends: Thank you for being part of this extraordinary journey. Your dedication, generosity, and belief in the power of education have made all the difference.

With deep appreciation and warm congratulations,

**Ann '75 and John '73 '74 Doerr**

Reginald DesRoches,
Rice University
President, 2025

"

Engineering is more than a discipline; it's a mindset, a way of thinking that solves complex problems and improves lives.

# ↗ Introduction

**Luay Nakhleh**
William and Stephanie Sick Dean
George R. Brown School of Engineering and Computing
Professor of Computer Science and of BioSciences

I don't remember the exact date, but I vividly recall the moment in spring 2004 when I received an email from Professor Keith Cooper, then chair of the Department of Computer Science, offering me a tenure-track assistant professor position at Rice University. I accepted immediately. I already knew Rice was where I wanted to be. I was drawn by its academic excellence and the strength of its research and teaching—but it was the "feel good" experience of the interview that sealed it for me.

That feeling—that intangible but unmistakable sense that this place is different—has never left me. And it's not just me. I hear it again and again from students, faculty, alumni, and collaborators. I hear it from parents, too. On the first day of the fall 2024 semester, President Reginald DesRoches drove a golf cart across campus, giving students rides to class while chatting and posing for photos along the way. One parent commented on social media, "Just another reason to love Rice." It's a small moment, but it speaks volumes about this special place.

That same spirit—of care, community, and purpose—runs through every part of Rice, and finds one of its most enduring expressions in the George R. Brown School of Engineering and Computing. As we celebrate more than a century of innovation—and mark the fiftieth anniversary of the school's naming—this moment is more than a chance to honor the past. It's a moment to reflect on who we are, what we value, and why this work matters.

## Our Culture Is Our Foundation

In the George R. Brown School of Engineering and Computing, we lead with people. We always have. Long before we were a nationally ranked school, before we led innovations in teaching and breakthroughs in research, we were a community guided by purpose. From the beginning, our school attracted individuals who weren't just looking to build things—they were looking to build a better world. That spirit continues to define us.

Our culture is rigorous but not rigid, collaborative but not complacent, and intellectually bold but humble. Our students and faculty approach their work with deep care, creativity, and a constant awareness of how their decisions affect others.

Leadership is also part of the school's DNA. Our students lead design teams developing solutions for real-world problems even before they graduate. Our faculty influence national conversations on energy, policy, and technology. Our alumni embody the mission of the school all over the world. They lead renewable energy companies pioneering sustainable solutions, innovate in healthcare by developing life-saving technologies, influence public policy to promote equity and access, and build start-ups tackling society's toughest challenges. Their work reflects the values they carried from Rice into the world. These remarkable contributions are not just markers of success—they are the ongoing expression of a community dedicated to making a meaningful difference. Rice leaders—whether in labs, boardrooms, or classrooms—are known not only for what they accomplish but for why they do it. Their work, grounded in purpose, reflects a profound commitment to positive impact. We believe that the best leaders listen deeply, think critically, and act with humility. Time and again, that is exactly the kind of leadership Rice fosters—within our classrooms, our labs, and across the globe.

It is this culture of care and leadership with purpose that has come to fuel the school's innovation and led to what we now define as *solving for greater good*.

## Solving for Greater Good: Our Mission in Action

At the George R. Brown School of Engineering and Computing, our mission is clear: We are solving for greater good. This is not a tagline. It is a deeply held belief—that engineering and computing are not ends in themselves, but tools for improving the human condition.

You see this mission in action across every department and discipline. Our faculty and students are tackling some of the most urgent challenges of our time—with empathy, ingenuity, and a sense of responsibility that reaches far beyond our campus. We see this mission in the work of our bioengineers developing neonatal technologies to reduce infant mortality in Africa, in environmental engineers working to ensure access to clean water, and in the global reach of OpenStax, which provides free, high-quality textbooks to millions. We also see it in research advancing sustainable computing, next-generation materials, and neurotechnologies that improve quality of life. Across all disciplines, Rice engineers and computer scientists are driven by the desire to make an impact.

Solving for greater good means asking not just what we *can* do, but what we *should* do. It means designing technologies that reflect the values of the communities they serve. It means embedding ethics, equity, and sustainability into every step of the process. And it means preparing graduates who carry those values into their careers. Solving for greater good isn't just our mission. It's our story and who we are. And one constant in our story is innovation. We don't just imagine the future—we engineer it. In our school, innovation doesn't end with a patent or a publication. It's measured by the good that it brings into the world—a direct expression of our mission.

## Engineered at Rice, Proven in Houston, Built for the World

Although Rice is one of the world's top private universities, its mission has always had a public heart, and this has much to do with calling Houston home. We are proud to be based in Houston—a global city, a hub of industry, a crossroads of culture, and a place that mirrors the complexity of the world. It is also a city of extraordinary assets: the energy capital of the world, home to the largest and one of the most advanced medical complexes on the planet, the nerve center of human spaceflight at

NASA's Johnson Space Center, and the most diverse metropolis in the United States. Houston's industrial depth, demographic diversity, and vast scale make it a living lab where Rice engineers and computer scientists test, refine, and apply bold ideas.

Rice engineers have played a central role in shaping this city's growth—in energy, medicine, manufacturing, transportation, and beyond—and we continue to engage Houston not only as our home but also as a proving ground for innovation with global impact. And while our reach is global, our identity is undeniably Texan. We embody the boldness, resourcefulness, and grit that define the Lone Star State. I like to think of our school as the pride of Texas, and I mean it—not as a boast, but as a responsibility.

Whether in Houston or halfway around the globe, Rice engineers and computer scientists are bringing Texas-sized ambition to the world's most pressing challenges—and doing so with the purpose, care, and excellence that characterize this school.

## What This Book Represents

This commemorative book is a celebration, but it's more than a retrospective. It's a reflection of the values that have always shaped us, and a tribute to the people who brought those values to life. It's a record of ideas that challenged assumptions and changed outcomes. It's a glimpse into how we became who we are—not just by following trends, but by shaping them.

You'll find milestones here—moments of transformation, excellence, and invention. But just as important, you'll find stories: stories of teachers who inspired, teams who persevered, students who dared, and discoveries that started small and grew into something world changing.

Ultimately, it's the recognition of our community's unshakable belief that engineering and computing can and should contribute to the betterment of the world.

## Looking Ahead

Now, more than two decades after joining Rice, I am honored to serve as dean and to witness each day the extraordinary impact of the George R. Brown School of Engineering and Computing. I write this introduction as the current dean and on behalf of all who came before me. This book reflects the work of generations—the faculty, staff, students, alumni, donors, and friends who shaped our school. Together, this community has built more than an academic institution—they've built a purpose-driven school deeply committed to solving for greater good.

The world is better because of this school—because of its people, its values, and its enduring mission to harness engineering and computing to improve lives and meet society's greatest challenges. I have no doubt this school will continue to shape a better world through its innovations in research, teaching, and service.

I invite you to explore this commemorative book, which offers an abridged but meaningful glimpse into the history of our storied school—a school that has been a catalyst for change, a community of purpose, and a beacon for what engineering and computing can achieve for humanity.

↗ Pre-1975

# Engineers from the Beginning

**From the moment** Rice opened its doors in 1912, engineering was central to its mission. The George R. Brown School of Engineering at Rice University did not come into existence until 1975, sixty-three years after the doors of the William M. Rice Institute for the Advancement of Literature, Science and Art first opened. But the practice of educating the problem-solvers known as engineers was baked into the DNA of Rice from the very beginning.

William Marsh Rice endowed the institution bearing his name in 1891, envisioning the education of a professional class of can-doers, designers, and planners who would work hand in glove with city leaders to overcome barriers that stood in the way of what Houston wanted to become. The Rice Institute was there to serve the city. And what Houston needed was engineers.

The Mechanical Laboratory was one of the first three buildings designed and built for the Rice Institute in 1912. From its inception, it housed mechanical engineering, electrical engineering, chemistry, and physics laboratories, as well as architectural drafting rooms and a machine shop. The attached powerhouse in the back of the building served as Rice's sole supplier of utilities such as electricity, heat, and water, and its smokestack was designed with a facade meant to resemble an Italian bell tower called a campanile.

When the Rice Institute opened its doors, Houston had eclipsed San Antonio as the largest city in Texas, with a population of close to one hundred thousand. The discovery of oil in 1901 ninety miles east at the Spindletop oil field had transformed Houston from boomtown to boom city. The number of residents more

→ Thomas Gaylord, professor of electrical engineering, adds formulas to a classroom blackboard.

Assumptions
1) uniform resistivity
2) neglect minority carrier injection
s >> L
no surface leakage current
probe-surface contact is hemi-
spherical
Potentiometric
Voltage
Measurement
I
s1
s2
s3
of current
between potential electrodes

than doubled in the eleven years between the gusher at Spindletop and the start of Rice.

Houston was the hub that connected the region, a promised land touted by the city's chamber of commerce as the place "where 17 railroads meet the sea." The Port of Houston emerged as a major shipping center after the great hurricane of 1900, the deadliest storm in US history, decimated the nearby Port of Galveston. About a decade after that disaster, county voters approved the dredging of the Houston Ship Channel, which was completed in 1914.

But the flat landscape, the wetlands, and the coastal environment stood in the way of accommodating all the people moving to the area. Most parts of the city flooded easily, and Houston lacked sufficient roads, sewers, bridges, streetlights, telephone poles, railroad tracks, and other infrastructure. Engineers could solve these problems.

The first engineering faculty members at Rice were educators, first and foremost. They were tasked with giving talks to educate the public, and in 1920 the first Rice Engineering Show drew an enormous crowd of ten thousand visitors. At its first commencement ceremony, held June 12, 1916, the Rice Institute awarded six bachelor of science degrees to graduates majoring in one of the four engineering disciplines then offered: chemical, civil, electrical, or mechanical engineering.

It was a time when engineers weren't just solving technical problems—they were shaping the future of the city itself. Few embodied that spirit more fully than Lewis B. Ryon. He began teaching civil engineering and served as chair of the civil engineering department in 1917 and continued to do so for the next forty-one years. In 1924, his already renowned reputation prompted the Houston city council to appoint Ryon city planning engineer, placing him on the city planning commission.

Ryon embraced the city position with great zeal, as he was given the opportunity to put the ideas that he taught to work. In his first year as city planning engineer, he enlisted the help of Rice students to create a street plan for

Houston that attempted to address the already heavy traffic volume. Ryon proposed four types of streets and thoroughfares for the city, including its first loop.

After giving his life to teaching at Rice, Ryon and his wife, Mae E. Ryon, generously gifted their total estate, valued at $750,000 (about $8 million in 2025), to Rice for construction of the Ryon Engineering Lab, which opened in 1965. “We came here without anything,” Ryon noted. “What we have, we made here in Houston. It’s all a product of Rice, and our association with Rice.”

↑ A group of students observe a flow simulator in the Ryon Engineering Laboratory during the 1972 Majors Day preview.

← In 1940, students worked on the fractional distillation equipment in the chemical engineering laboratory.

## The Boomtown and the Builders

Rice Engineering’s scope broadened over the following decades, especially with the focus on national defense after the United States

An aerial view of the Rice campus, taken from the Texas Medical Center, shows Hermann Hospital in the foreground.

I'd walk into a large room, which was the computer. I don't mean the computer was *in the* room; the computer *was* the room.

**Bart Sinclair,** professor of electrical and computer engineering

entered World War II in 1941. Following the end of the war, wartime production wound down and the production of petrochemical products like plastic could be ramped up. Oil companies, manufacturers, and shippers scrambled to fund process-oriented research at Rice that could help them adapt to this shift in production. Houston grew into the petrochemical and energy capital of the world, with a booming industrial complex lining the banks of the Houston Ship Channel. The industry provided Rice graduates with job opportunities that set them up for leadership roles both at home and abroad.

At Rice, the pinnacle of this postwar boom was the completion of the Abercrombie Engineering Laboratory in 1948. Oilman J. S. "Jim" and Lillie Abercrombie, active philanthropists and parents of Rice alumna Josephine Abercrombie '46, donated $500,000 in 1947 toward construction of the new building. The lab was designed as an all-purpose research center, with a machine shop, a forge, a foundry, a metallurgy shop, welding and wood shops, and an internal combustion lab.

## The Machine Upstairs

Nearly ten years after the Abercrombie Engineering Laboratory was dedicated, the second floor of the building became the birthplace of the largest and fastest computer ever assembled on a university campus.

Construction of the Rice Institute Computer (R1) began in September 1956, and by 1961 the R1 was supporting research requiring calculations that would have been impossible without a large computer's processing power. It had a huge impact on Rice research throughout the 1960s—and it also had a huge physical footprint.

"It was an important tool for a lot of computation-oriented research that was going on here," explained Bart Sinclair, professor of electrical and computer engineering and Rice alum. "Up on the second floor [of Abercrombie], I'd walk into a large room, which was the computer. I don't mean the computer was *in the* room; the computer *was* the room."

“It was the start of a long and very productive association of computer technology hardware and software with Rice University,” Sinclair said. The first computer at Rice, a Litton LGP-30, had arrived in 1957. But the R1 was bigger and faster—and completely designed and constructed at Rice by faculty, staff, and graduate students.

Thanks to the R1, Rice was able to offer programs in software and hardware design that attracted faculty with backgrounds in computer science and computer engineering. The R1 was also a factor behind the Rice Institute naming LeVan Griffis as its first dean of engineering in 1959. Griffis believed that the R1 would be “the basis for new computers,” leading to “revolutionary new concepts in the same field.”

With the expansion of its educational offerings and enrollment, the Rice Institute changed its name in 1960 to William Marsh Rice University to reflect its significant growth.

## Signals and Breakthroughs

Electrical engineering was undergoing a paradigm shift too. Rice had been on the forefront of research and education in digital signal processing (DSP), a fundamental part of many technologies, from its inception. Electrical and computer engineering professors Tom W. Parks and C. Sidney Burrus began their DSP research in the late 1960s and offered the first Rice course on DSP in 1968.

In 1972, Parks and his student Jim McClellan published a paper on finite impulse response

↑ Lead programmer John Iliffe stands before the Arithmetic Unit Racks of the R1 computer, Rice's first large-scale research computer.

← Philanthropist J.S. Abercrombie, family, and friends admire the sculpture on the facade of Rice Institute's Abercrombie Engineering Laboratory.

By breaking bar
and building co
these pioneers
establish the fo
of the future.

riers
mmunity,
helped
undation

## Anna Rebecca Turner

In 1929, Anna Rebecca Turner (above) became the first woman to graduate from Rice Engineering. She defied female dress codes daily, wearing coveralls so that she could work safely in the labs. Her pioneering spirit led her to become the first female engineering graduate in Texas. Close behind her was Margaret Hutchinson Rousseau, who graduated from Rice with a chemical engineering degree in 1932. Rousseau went on to study at Massachusetts Institute of Technology and, in 1937, became the first woman to earn a doctorate in chemical engineering in the United States. She left her mark on the engineering world by designing the first commercial penicillin production plant.

## Rice Engineering Alumni

Rice Engineering Alumni (REA) supports, honors, and connects Rice engineers before and after graduation. Formed in 1938, the REA is the oldest alumni affinity group at Rice and is dedicated to providing support to the George R. Brown School of Engineering and Computing. REA provides over $150,000 in awards and scholarships to outstanding students each year, recruits students for meaningful internships, and facilitates opportunities for alumni to interact with Rice students.

# ↗ The Deans List

## LeVan Griffis
(1959–1960)

LeVan Griffis expanded Rice Engineering's reach by strengthening applied research, building industry partnerships, and championing a broad, fundamentals-based engineering education. He deepened ties with national engineering societies, supported the growth of graduate programs, and encouraged interdisciplinary collaboration. His leadership helped shape the school's research mission and guided its transition as the Rice Institute became William Marsh Rice University in 1960.

**LeVan Griffis**

## Franz Brotzen
(1961–1966)

An internationally recognized lecturer and mentor, Franz Brotzen helped shape Rice into a more research-focused university in the 1960s and founded its materials science program. His influence extended beyond the classroom, inspiring generations of students to pursue their interests abroad. He received a Guggenheim Fellowship and Rice's George R. Brown Award for Excellence in Teaching four times during his fifty-five-year career at Rice.

**Franz Brotzen**

## William Gordon
(1967–1974)

William Gordon brought international acclaim to Rice through his transformative leadership and technical vision. After leading the creation of the Arecibo Ionospheric Observatory at Cornell University, Gordon joined Rice to further the research ambitions of its president, Kenneth Pitzer. As dean and later provost and vice president, Gordon championed interdisciplinary collaboration and strengthened the curriculum, helping position Rice as a top-tier research university. He was elected to the National Academy of Sciences and to the National Academy of Engineering, among others.

**William Gordon**

(FIR) filter design. This would become a breakthrough in DSP, as these filters could be efficiently implemented using a fast Fourier transform (FFT). With the Parks-McClellan method, "You could do [digital filtering] on a little computer in a few minutes instead of on giant computers in several days. Before that, it was just impossible," explained Burrus.

With this greater efficiency and practicality, early applications of DSP included critical fields such as defense, oil exploration, and medicine. DSP would become the engine powering Rice Engineering's reputation for being at the cutting edge of research.

## Engineering the Space Race

Rice began research collaborations with the National Aeronautics and Space Administration in 1959, months after the agency was founded. But starting as early as 1958, former Rice Engineering student and Rice board chairman George R. Brown was hard at work behind the scenes with his former Rice roommate, Congressman Albert Thomas '20, ensuring that Houston and Rice would play leading roles in the space race. At Brown's urging, Humble Oil donated 1,020 acres of land to Rice University, which then offered the land to the National Space Council.

In 1961, NASA officials toured Houston and met top Rice faculty at the home of Franz Brotzen, then Rice's dean of engineering. That same year, the donated land was chosen as the site of the NASA Manned Spacecraft Center (now Johnson Space Center). Houston's reputation as "Space City" was preserved in history on September 12, 1962, when US President John F. Kennedy gave his famous "We choose to go to the moon" speech at Rice University Stadium.

The arrival of William Gordon at Rice in 1966 as the first dean of science and engineering, and as a faculty member of the space science and

↓ On September 12, 1962, as part of Rice's semicentennial celebrations, then-president John F. Kennedy spoke at Rice Stadium, challenging the United States to become "the world's greatest space-faring nation."

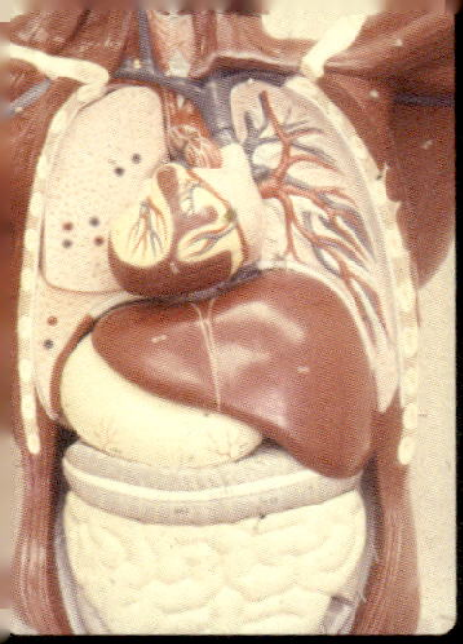

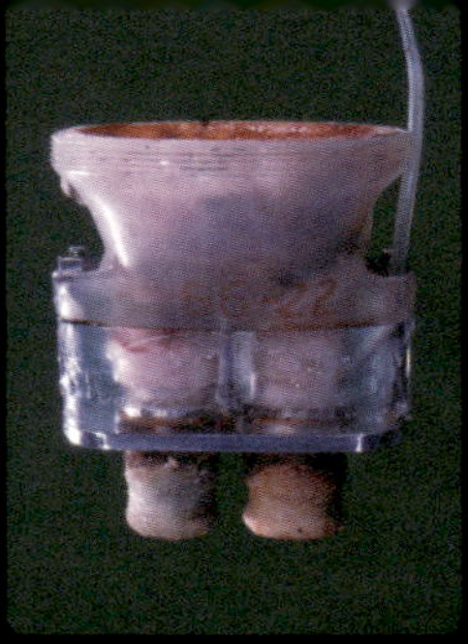
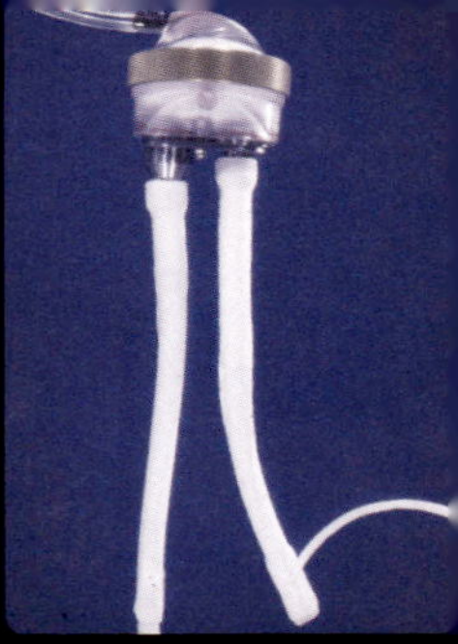
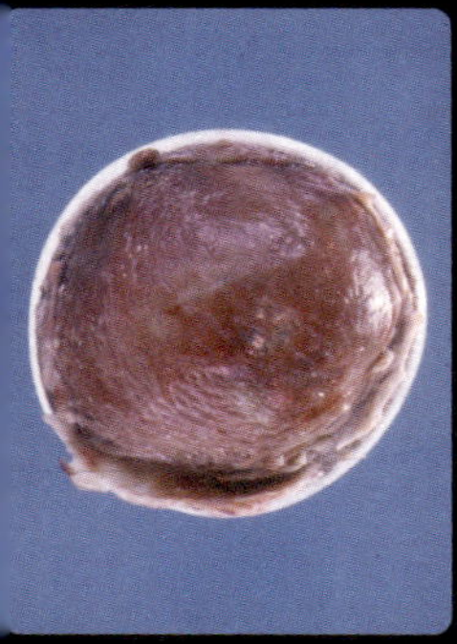

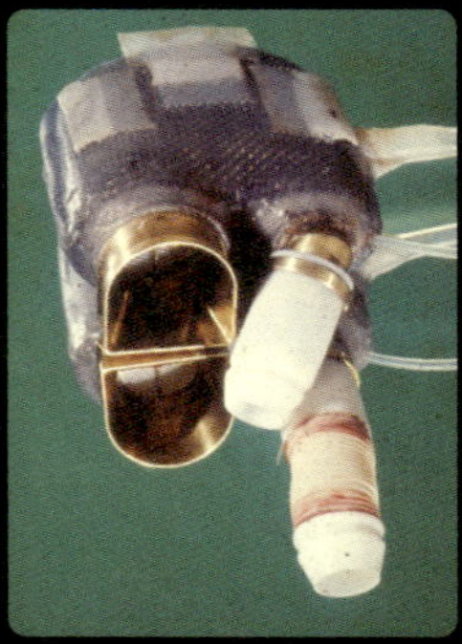

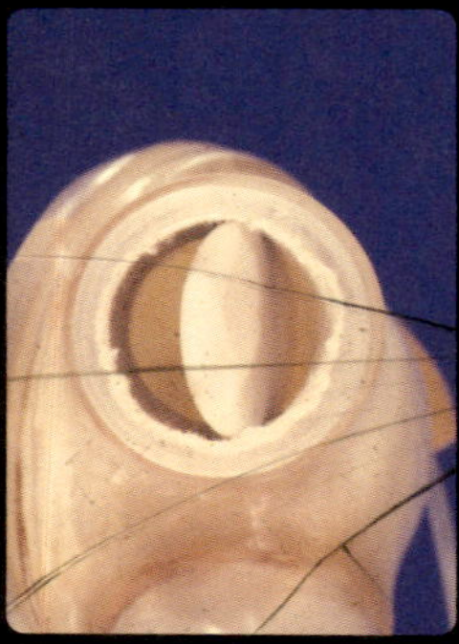
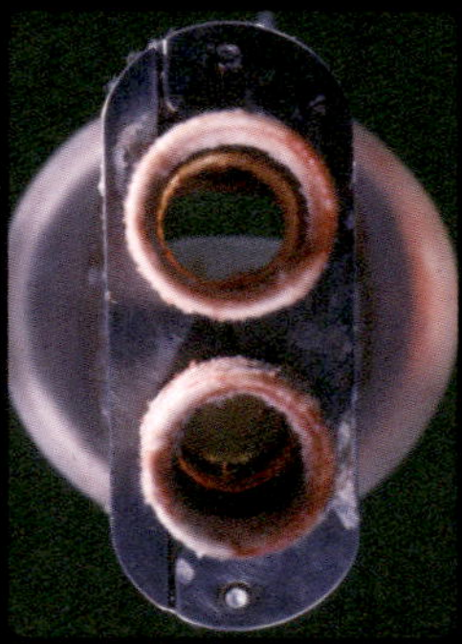
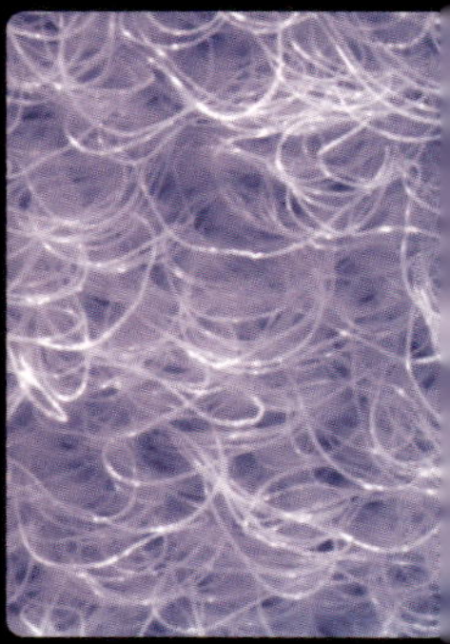
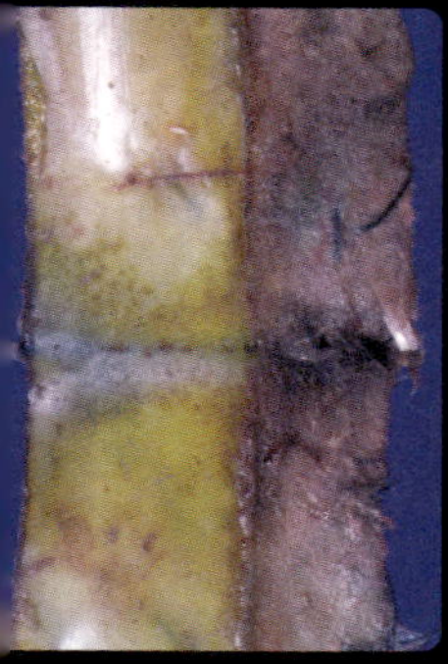
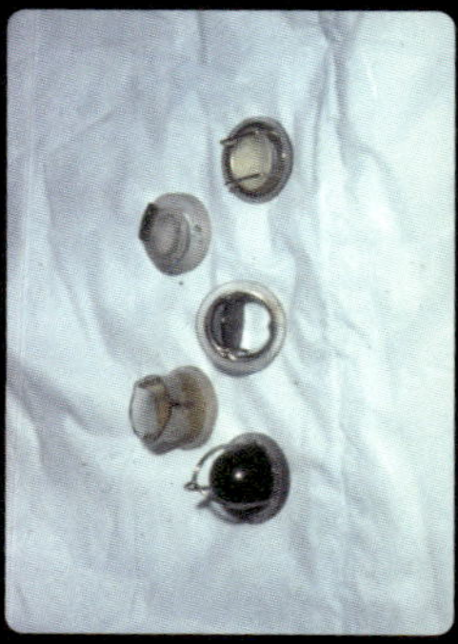
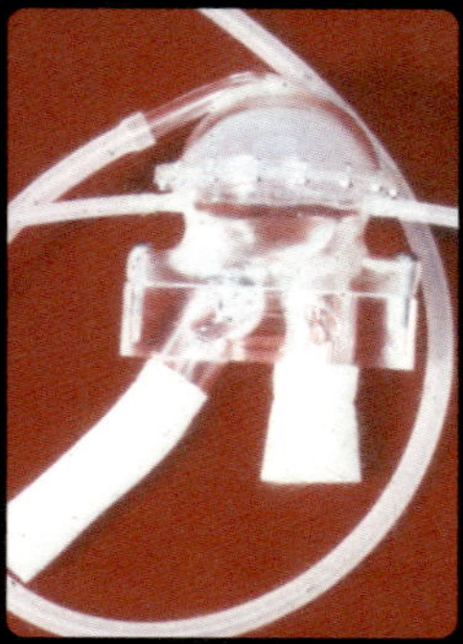
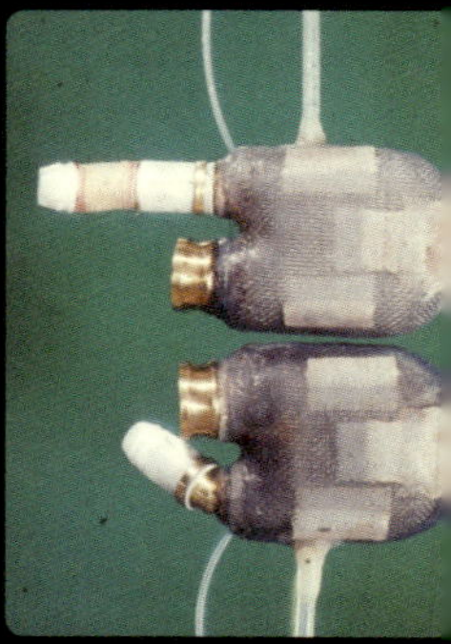
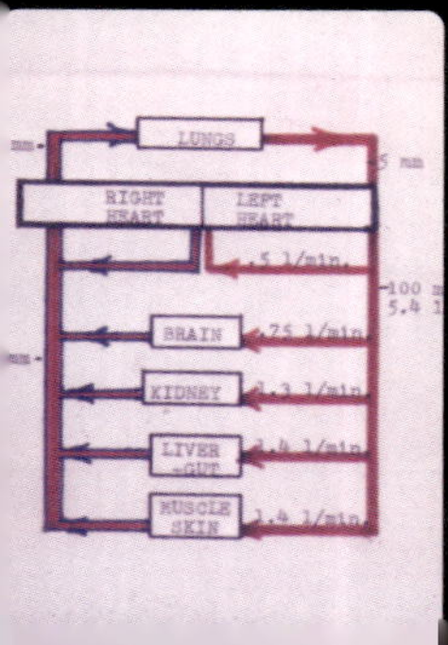
LUNGS
RIGHT HEART
LEFT HEART
BRAIN
KIDNEY
LIVER -GUT
MUSCLE SKIN
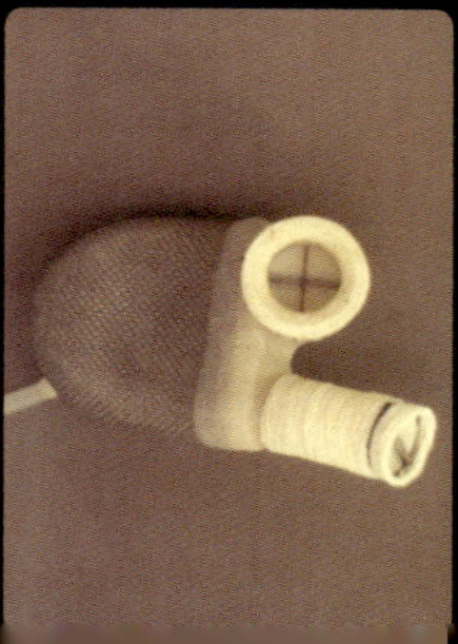
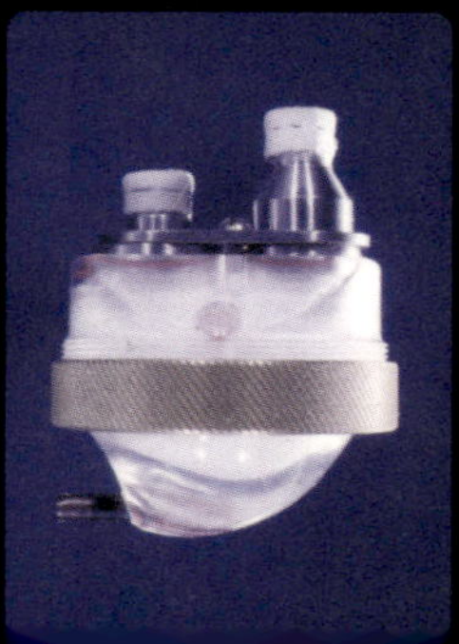
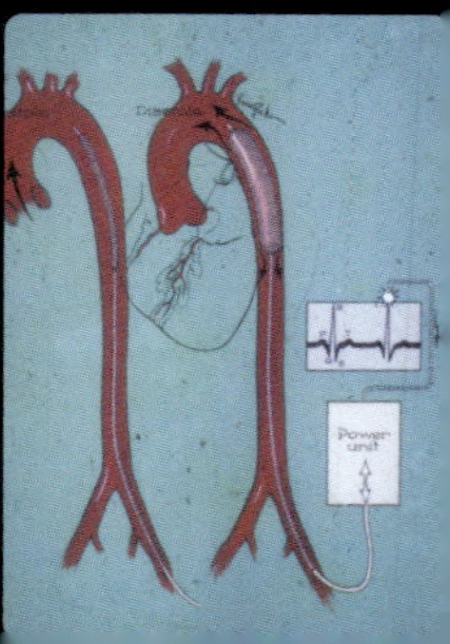

↗ Professor William W. Akers conducts pioneering research on the development of an artificial heart.

← The components of the artificial heart highlight the engineering precision and innovation behind this lifesaving device.

electrical engineering departments, signified Rice's embrace of outer space as a new frontier for research. Gordon had previously built and operated the Arecibo telescope in Puerto Rico. It was the largest single-dish radio telescope built at that time. Gordon sent Rice graduate students to work on the telescope, a protocol that would continue until his retirement in 1985.

## Hearts and Minds

In 1964, the Laboratory for Biomedical Engineering was established by William W. Akers, chair of the Department of Chemical Engineering, in the Abercrombie Engineering Laboratory. That same year, Akers, J. David Hellums, and other Rice engineers began collaborating with Dr. Michael DeBakey at the Baylor College of Medicine. DeBakey needed engineers—and biologists, chemists, and mathematicians—to turn his research into something real, an artificial heart.

Akers led the team of twenty-three Rice engineers tasked with developing engineering designs for a mechanical pump that could maintain circulation in a patient beyond four days, and Hellums contributed his expertise in blood flow and platelet behaviors. The Baylor team, led by DeBakey, was responsible for lab testing and solving the physiological problems of implantation.

In 1965, the team produced a masterpiece. Akers later explained, "The heart is two pumps. The left ventricle pumps blood from the lungs through the entire body, which flows back to the heart through the veins. The right ventricle then pumps the blood back through the lungs. The left ventricle is the most critical." In 1966, the Rice-built left ventricular bypass pump was successfully implanted in a patient, keeping her alive during a ten-day recovery from heart surgery until her own heart was strong enough to work on its own.

Ultimately, the full artificial heart device was not a viable long-term solution, but the project's research laid the groundwork for future breakthroughs in cardiovascular assist devices.

"The sixties were a time of great imagination, when people thought science could perform miracles—artificial hearts and brains, cities under the sea," Akers said some years later.

The joint effort also marked the beginning of a long and deep relationship between

D
B

> We try to stress the *why* rather than the *how*.
>
> **Arthur Busch,** professor of environmental engineering

the university and the nearby Texas Medical Center, which would grow into the largest medical complex in the world.

### The Birth of Environmental Engineering

In 1967, environmental engineering professor Arthur Busch wrote an essay titled "The Rice University Approach to Graduate Study, and the Research in Environmental Science and Engineering" for the journal *Rice University Studies*, detailing the genesis of environmental science and engineering at Rice.

Busch was regarded as a leading light in the nascent environmental movement, and he played an influential role in recruiting talented people to the faculty. One of these recruits, in 1966, was a self-described "microbiologist parading as an environmental engineer" named C. H. "Herb" Ward, who went on to chair the Department of Environmental Science and Engineering for twenty-two years.

The department set out to teach students to anticipate global problems. The need to mitigate impacts of pollution was a huge driver. Contamination of land, air, and water had become a worldwide problem. Solutions—and engineers—were needed badly.

Rice Engineering was just getting started.

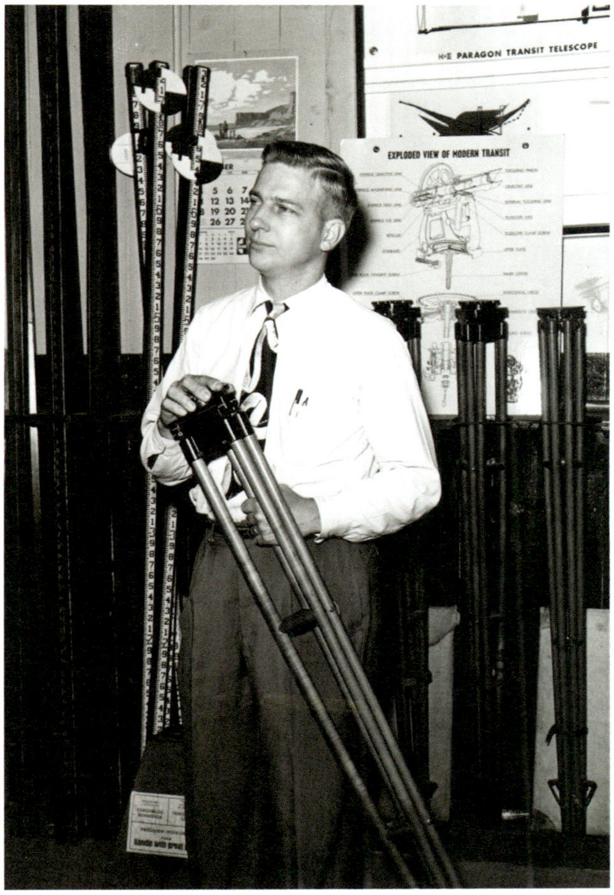

Rice teacher named to EPA post

ARTHUR W. BUSCH

Outspoken Arthur W. Busch, the 45-year-old professor of environmental engineering at Rice University, has been appointed regional administrator of the Environmental Protection Agency.

EPA Administrator William D. Ruckelhaus made the announcement Monday.

Busch, a Republican who ran a strong but unsuccessful race to unseat Democratic U.S. Rep. Bob Casey last November, was the last of 10 regional administrators to be appointed. He will start work for the agency Jan. 1 and will have headquarters in Dallas.

An acknowledged pollution control expert, Busch will be responsible for EPA programs on air and water pollution control, noise abatement, solid waste disposal, pesticide and radiation regulation in Texas, Louisiana, Arkansas, New Mexico and Oklahoma.

Born in Houston, Busch was graduated with a degree in civil engineering from Texas Technological University and a master of science degree in sanitary engineering from Massachusetts Institute of Technology in 1952.

Since then he has served as a teacher at Rice and a consultant on industrial pollution in the Houston area. He has won several professional awards and citations for his work in pollution control and research.

↑↑ Professor Art Busch handles equipment in the Engineering Annex Building.

↑ A 1971 newspaper article discusses Art Busch's appointment to the Environmental Protection Agency.

← In 1964, Professor Art Busch conducts research in the mechanical engineering lab.

# Breaking Ground

**1** Members of the 1925-1926 Rice Engineering Society gather for a group photo.

**2** In 1912, construction on the Mechanical Engineering Laboratory and the Campanile moved forward.

**3** John Cox speaks during the groundbreaking of the Mechanical Engineering Building in 1984.

**4** Norman Hackerman and John Cox (center and right) celebrate the groundbreaking with guests.

From turning soil to turning ideas into impact, Rice engineers have always broken new ground. In 1912, the first shovels struck earth to build the Rice Institute, and soon after, its engineers began shaping Houston's future. By 1920, ten thousand Houstonians flocked to the inaugural Engineering Show to witness marvels like remote-controlled cars and combustion engines, proof that innovation was in Rice's DNA. Across the decades, Rice engineers have

**1**

2

3

4

1

2

3

1

2

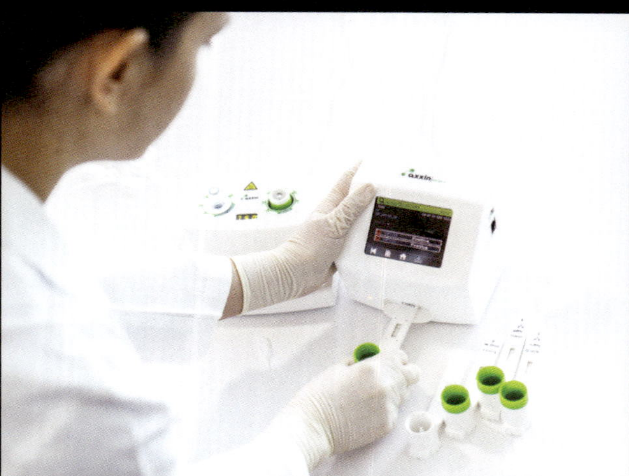

3

4

5

**1** An interdisciplinary team at Rice led by Lauren Stadler, Kathy Ensor, and Loren Hopkins worked with the Houston Health Department to analyze wastewater samples in the lab to stay ahead of potential measles outbreaks.

**2** Rice bioengineers in Rebecca Richards-Kortum's lab developed a low-cost, point-of-care DNA test for HPV that could expand cervical cancer screening access in low- and middle-income countries.

**3** Matteo Pasquali, director of Rice's Carbon Hub, holds carbon nanotube fibers that can be fully recycled without losing structure or performance, offering a sustainable alternative to metals, polymers, and conventional carbon fibers.

**4** A $FAPbI_3$ solar cell solution is enhanced with specially designed 2D perovskites, improving both efficiency and long-term stability.

**5** Rice University engineers in the lab of Jacob Robinson and collaborators have developed the first miniaturized brain stimulator shown to work in a human patient.

**6** In 2011, President Barack Obama awarded Richard Tapia the National Medal of Science, the highest honor given by the US government to a scientist or engineer.

6

1

2

3

5228A

# ↗ 1975–1984

# A School of Engineering Is Born

**On November 12, 1975**, the George R. Brown School of Engineering was officially dedicated, consisting of five departments: chemical engineering, civil engineering, electrical engineering, environmental science and engineering, and mechanical engineering and materials science. Alan Chapman, who had graduated from the Rice Institute in 1945, was named the first dean of the George R. Brown School of Engineering upon its dedication.

## The Man Behind the Name

George R. Brown was an engineering student who had left Rice for Colorado before graduation, but he was not gone for long. He returned to Houston to join his family's construction business, Brown & Root. Brown went on to have a tremendous impact on Rice and Houston as both a benefactor and a leader.

Brown joined the Rice board of trustees in 1943 and became the first former Rice student to be elected board chair in 1950, a position he held for the next fifteen years. He guided construction of the Fondren Library, Anderson Hall, and the Abercrombie Engineering Laboratory, among many other projects.

He was a hands-on trustee. "George Brown walked this campus at least once a month with a yellow pad and a pen—*Oh, there's a cracked window on the northwest corner of the physics building*," former Rice historian Melissa Kean said. "And he knew the faculty."

Brown played a key role in Rice's $33-million fundraising campaign from 1962 to 1968, personally pledging $4.5 million and contributing another $3.5 million through the Brown Foundation, Inc. In 1974 and 1975, the foundation gave an additional $1 million to develop fine

→ George R. Brown speaks to a gathering of benefactors and students at Rice University.

↑ A historic photo shows George R. Brown speaking to the Rice community.

→ Herb Ward (fourth from the right) gathers with colleagues beside the granite monoliths in the Engineering Quadrangle.

arts at Rice and $750,000 to establish a chair in engineering named for Jim Abercrombie.

In 1975, nearly sixty years after leaving Rice as a student, Brown was greatly honored when the university dedicated the George R. Brown School of Engineering in his name.

## Making Room to Grow

The creation of the George R. Brown School of Engineering was in part due to the volume and breadth of research being done at the university. There were more researchers than the existing facilities could accommodate. This situation got the attention of Rice president Norman Hackerman, who called for the creation of two schools—the School of Natural Sciences and the School of Engineering—to meet the demand.

The prospect of collaboration attracted Kyriacos Zygourakis, a chemical and biomolecular engineer, to join the Rice faculty in 1980. Rice's close-knit community fostered interdisciplinary teamwork, a marked contrast to what Zygourakis had observed elsewhere: "I have heard of so many schools where different departments don't talk to each other. This wasn't the case here. I came here with the mathematical and computational tools. I found people who were really interested in this, and at the same time, I found people who actually needed these tools. This started my collaboration with the biomedical engineers and chemical engineering."

The collaborative idea expanded beyond the walls of the campus in 1982 with the creation of the Houston Area Research Center, a four-university consortium with Rice, the University of Houston, the University of Texas, and Texas A&M University based in The Woodlands, a planned community just north of Houston.

In 1980, J. David Hellums succeeded Alan Chapman as dean of the School of Engineering, helping guide the school through a period of growing research strength and institutional ambition.

## Cleaning Up

The rise of environmental and ecological awareness in the 1970s, coupled with concern over the impacts of pollution and toxic waste, was raising the profile of the Department of Environmental Science and Engineering, where faculty researchers had been looking for answers for some years. Several Rice engineers were recruited to work on large national projects.

Herb Ward, professor of civil and environmental engineering, and of ecology and evolutionary biology, played a major role in devising solutions to groundwater contamination across the United States. In 1979, he was appointed director of the National Center for Groundwater Research at Rice, a position he would hold for eighteen years.

Ward's collaborator, Philip Bedient, Herman Brown Professor of Engineering, came to Rice in 1975 to work on The Woodlands planned community, which was receiving funding from the Environmental Protection Agency. "The project to help design a new town or a new city with nature-based solutions was very exciting," Bedient said.

Ward's role directing national research in groundwater influenced Bedient's subsequent work in groundwater contamination and cleanup and analysis. "There were major environmental contamination problems all over the United States in water and air, in lakes and rivers. Rivers were catching fire. It was an awakening," Bedient explained. "The environmental movement had launched in 1970, and lots of funding was provided for graduate student training as well as for performing research. The funding just kept flowing and the problems kept getting more and more complex."

> The funding just kept flowing and the problems kept getting more and more complex.
>
> **Philip Bedient,**
> Herman Brown
> Professor of Engineering

### Designing with a Pulse

Innovations emerging from the Department of Chemical Engineering and the associated Biomedical Engineering Laboratory boosted Rice's scientific profile and strengthened its ties to the broader medical community as faculty and students developed artificial heart components, improved heart-valve designs, and pioneered polymer-based drug delivery systems.

A moral compass accompanied the breakthroughs. In 1977, William Akers had noted that, in the quest to develop an artificial heart, "there were

In 1984, Rice dedicated the Mechanical Engineering Building, located between the Ryon Engineering Laboratory and the Chemistry Building.

MECHANICAL ENGINEERING

"

# The pursuit of excellence equity continu inspire genera of engineers computer scie

and
es to
tions
and
ntists.

**Richard Tapia**

Richard Tapia (second from left), University Professor and Maxfield-Oshman Chair in Engineering, is an internationally recognized advocate for inclusivity in STEM and a pioneer in computational and mathematical sciences. Tapia, whose parents emigrated from Mexico, broke barriers in academia, becoming the first Hispanic elected to the National Academy of Engineering. His research in optimization theory and numerical analysis has advanced fields from energy to aerospace. Equally impactful is his lifelong commitment to increasing representation in science and engineering. Through mentoring, outreach, and programming through the Tapia Center for Excellence and Equity in Education, he has inspired generations of students. In recognition of his contributions, Tapia was presented with the National Medal of Science in 2011 by President Barack Obama. He has received numerous other honors, including election to the American Academy of Arts and Sciences and receiving the National Science Board's Vannevar Bush Award.

# ↗ The Deans List

## Alan Chapman (1975–1979)

Alan Chapman, the founding dean of the George R. Brown School of Engineering, is remembered for his leadership, scholarship, and dedication to teaching. He chaired the mechanical engineering department three times, served as vice president for administration, and won Rice's George R. Brown Award for Superior Teaching five times. In 1984, he received the university's highest teaching honor, the George R. Brown Award for Excellence in Teaching. Though he retired in 1995, he continued teaching until 2006. An expert in heat transfer and thermodynamics, Chapman authored twelve books and conducted research critical to the development of helmet visors worn by NASA's Apollo astronauts, leaving a lasting mark on engineering education and space exploration.

Alan Chapman

## J. David Hellums (1980–1987)

J. David Hellums, a biomedical researcher and former dean of engineering, played a foundational role in establishing bioengineering at Rice. His involvement in the artificial heart project and in developing the bioengineering lab within the chemical engineering department ultimately led to the formation of the Department of Bioengineering, where he served as acting chair. Elected to the National Academy of Engineering, Hellums was a leader in cardiac assist devices and red blood cell mechanics. His groundbreaking work on blood flow, vascular physiology, and thrombosis advanced understanding of human disease and enabled new therapies. He had a lasting impact on healthcare innovation and Rice's standing in bioengineering education and research.

J. David Hellums

as many debates on campus concerning the ethics of an artificial heart as its mechanical feasibility." To address these issues, Rice, Baylor College of Medicine, and the Institute of Religion of the Texas Medical Center formed the Center for Ethics, Medicine, and Public Issues, which created joint graduate programs between Rice and the UT School of Medicine.

## A Bigger Tent

In 1984, the Mechanical Engineering Building, located between the Ryon Engineering Laboratory and the Chemistry Building, was dedicated. Previously, the offices and labs for mechanical engineering had been spread out among spaces in the Abercrombie Engineering Laboratory, the Ryon Engineering Laboratory, and the Space Science and Technology Building. Thanks to a generous gift of $3.5 million from John Cox '45, a trustee and benefactor, the Department of Mechanical Engineering and Materials Science was now housed in a single structure.

At the building's dedication, Alan Chapman, the former dean of engineering, described the coming academic year as "the most significant point in the history of the department."

## A Defining Realignment

The first recognition of computer science as an academic discipline at Rice came in 1964, with the naming of Rice Engineering alumnus Henry Rachford, Jr., '45 as professor of computer science and chair of the university's Committee on Computers.

In 1971, Stuart Lynn was appointed the founding director of the Institute for Computer Services and Applications (ICSA), which was conceived as a means of providing information-processing services to the Rice community and exploring the role of computers in society at large. For several years, starting in 1972, ICSA processed Houston's election data and served it to the public media for news broadcasts. Key faculty from the electrical engineering and mathematical sciences departments guided ICSA's development, including Robert Jump, Edward Fuestel, Ken Kennedy, and Robert "Corky" Cartwright. In 1982, Jump, Kennedy, and Cartwright joined research professor John E. Dennis in applying for and outcompeting computer science departments from other universities to win a National Science Foundation community-engaged research grant. This grant, for $1.5 million spread over five years, added to the momentum that led to the creation of the Department of Computer Science in 1984.

The rise of computer science had a transformative impact on Rice, fundamentally altering its academic structure, research capacity, and technological infrastructure. Computer science had parity with other established academic disciplines within the School of Engineering, reflecting the increasing importance of computing within traditional engineering fields.

The creation of the Department of Computer Science also signaled a broader realignment within the university—one that prompted other disciplines, like mathematical sciences, to consider where they naturally belonged. Mathematical sciences belonged in engineering, said Richard Tapia, who chaired the Department of Mathematical Sciences from 1978 to 1983: "We were already acting as if we were in the engineering school. In many ways, we were viewed as an engineering department all along. It didn't change much in our day-to-day activity, but it did change much in the way we saw the building and the future. Working with the engineering school, everybody realized we just fit there. We asked the provost to move us from natural sciences to engineering, which would be our natural home."

That request was approved: In 1984, the mathematical sciences and computer science departments were incorporated into the George R. Brown School of Engineering.

The scope of engineering in general was broadening at a dizzying pace. Adding departments that played major roles in collaborative projects kept Rice Engineering on the cutting edge. With the creation of prestigious endowed chairs—including the J. S. Abercrombie Chair in Engineering, the Harry S. Cameron Chair in Mechanical Engineering, and the Stanley C. Moore Chair in Engineering—Rice had matured into a top-shelf research institution, with the support to recruit and retain world-class faculty.

# Beyond the Books

→ Electrical engineering student Frederick Foston (right) plays chess with another student in a dorm room.

When alumni look back on their student days, it's the people and shared moments that made Rice feel like home. Beyond the classroom, the memories live in late-night conversations during study breaks, midday walks through the campus arboretum, and weekends spent exploring Houston. These experiences are what make the Rice journey unique. They shape identities, spark lifelong friendships, and create memories held close for years to come. At Rice, what happens beyond the books is just as important as what happens within them.

FOSTON

55Do 18-A 09-198-1.89

1

2

3

4

5

6

**1** Students enjoy an ice cream social outside Duncan Hall to kick off the academic year.

**2** Ted Adams, a member of the Rice Men's Soccer Club, later served as president of the Rice Engineering Alumni Association.

**3** Students enjoy lunch with William "Dr. Bill" Wilson, who taught in the Department of Electrical and Computer Engineering for thirty-four years.

**4** Faculty and students celebrate Holi, the Hindu

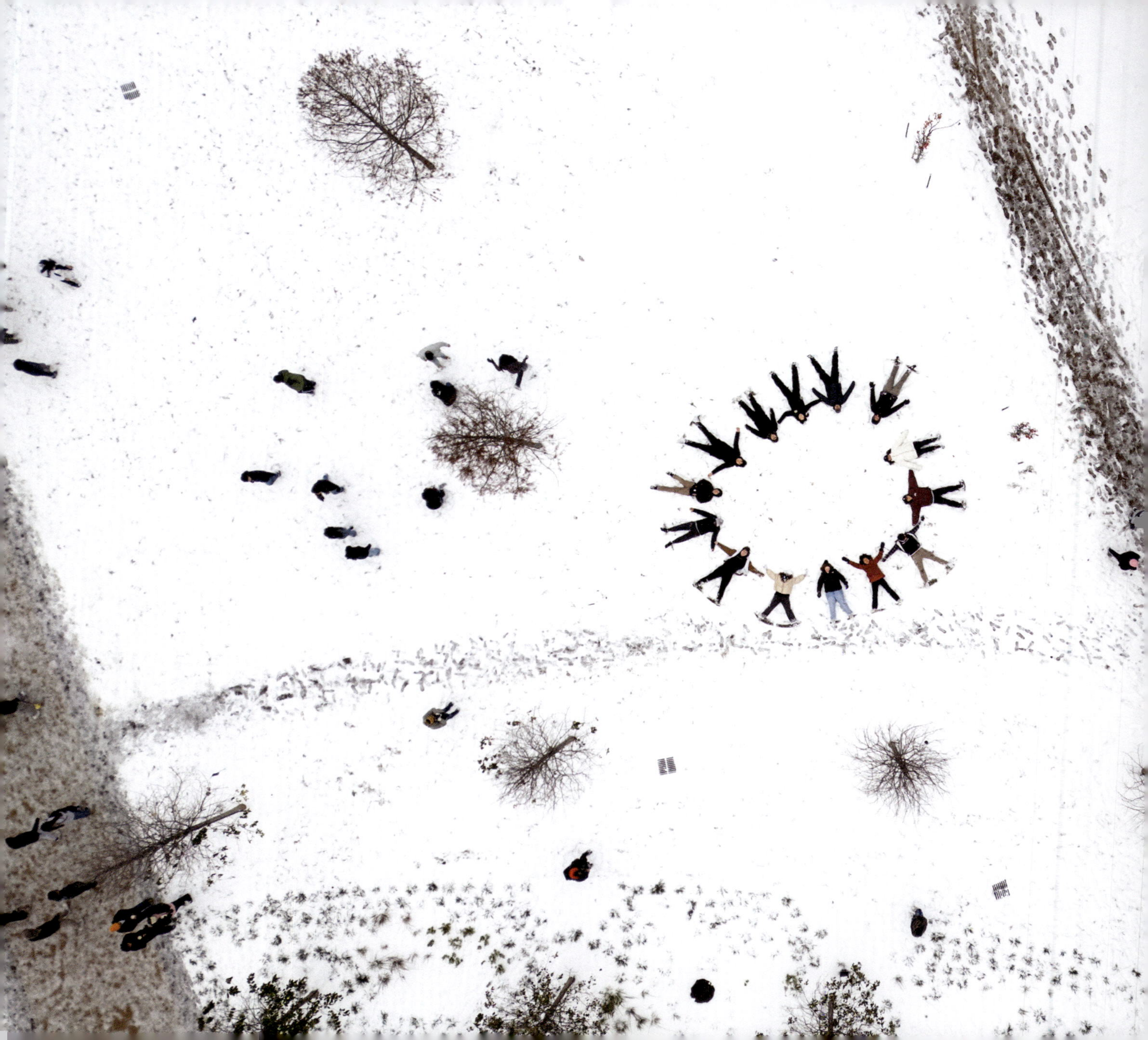

Rice University experienced a rare and historic snowfall on January 21, 2025, transforming the campus into a winter wonderland. The snowfall, an unusual event for Houston, blanketed the university's signature live oaks, the arches of Lovett Hall, and the Academic Quadrangle. Students enjoyed the weather, engaging in snowball fights, building snow sculptures, and making memories against the backdrop of the wintry campus.

Apple Macintosh
Tag: 34196

# ↗ 1985–1994

# From Heart Pumps to Healing Cells

**The early evolution** of bioengineering at Rice University was driven by researchers and academic leaders whose work fostered an unprecedented level of interdisciplinary collaboration.

Joint projects with the Texas Medical Center were initially at the individual level. The Cox Laboratory for Biomedical Engineering, created in 1964, was among the first structures that facilitated interaction and collaboration at the institutional level between Rice faculty and researchers at the Texas Medical Center.

The Institute of Biosciences and Bioengineering, established at Rice in 1986 to strengthen global research and to support faculty in interdisciplinary research and education, marked a turning point. Engineers, biologists, and medical researchers could work together in shared laboratories. At the time, few universities saw the possibilities of merging engineering and science with the biosciences. And the institute's centralized location near the Texas Medical Center was a key component of its success.

That atmosphere attracted additional talent and funding. It was during this period of growth that Michael M. Carroll was appointed dean of engineering in 1988, ushering in a decade of emerging research areas and expanded academic leadership.

Larry McIntire, E. D. Butcher Professor of Bioengineering and Chemical Engineering, emerged as a leader in promoting interdisciplinary research, inspired by the work of William Akers and J. David Hellums on Michael DeBakey's artificial heart project. As institute chair, McIntire championed numerous projects, including the investigation of cell replacement therapies, the development of artificial blood, and protein engineering. Studies in genetically

→ Students collaborate in the Biomedical Engineering Lab.

engineered proteins, using bacteria to produce therapeutic proteins, opened the door to potential breakthroughs in treating heart attacks, viral diseases, and cancers. A pilot plant facility within the institute supported research in bioprocessing and functioned as an incubator for start-ups.

The Texas Medical Center wasn't the only entity invested in bioengineering at Rice. NASA also called on Rice bioengineering researchers to collaborate in investigating the effect of mechanical forces on cells and tissues, and how the lack of mechanical forces or microgravity can affect tissue physiology.

Rice's bioengineering roots in chemical engineering made a distinctive difference. "[Most] bioengineering or biomedical engineering departments evolved out of electrical engineering or mechanical engineering," said Antonios Mikos, a tissue engineer who joined the Rice faculty in 1992.

Convergence science—the integration of methods from diverse disciplines—required collaboration among colleagues.

"People recognized that to solve big problems like diabetes and cancer, you couldn't rely on a single investigator," Mikos explained. "You need to collaborate with people outside your field of expertise."

## The Advent of Data Science

Statistics had become a department in the School of Social Sciences in 1987, when it separated from the Department of Mathematical Sciences. Its first faculty hire was Kathy Ensor. Three years later, due to increasing demand for data, statistics became part of the George R. Brown School of Engineering.

"The faculty in the School of Social Sciences strongly supported the creation of the department, and we still have exceptionally strong ties with them," Ensor said. "Our department was notably one of the founding international drivers of what has become the discipline of data science. And so it really made sense for us to be homed in engineering."

James R. Thompson, Noah Harding Emeritus Professor of Statistics, was the founding department chair and a nationally recognized expert in applying advanced statistical methods to real-world problems. Notably, he authored thirteen fundamental books that are still influential in the field today. According to Ensor, "His work in cancer modeling set the stage for a deep understanding of how metastatic breast cancer progressed, establishing how best to quantify the disease through all stages."

Founding members of the Department of Statistics brought deep expertise and a spirit of innovation that shaped the department's early direction. Rice alumnus David Scott, an international leader in multivariate methods, helped establish its reputation for excellence in data science. Marek Kimmel contributed foundational work at the intersection of genomics, personalized

↑ Mechanical engineering students stand with the Ricobot I, a robot built in the department's laboratories.

↗ Robotic hands developed in Rice's mechanical engineering labs demonstrate how sensory feedback improves grasping techniques.

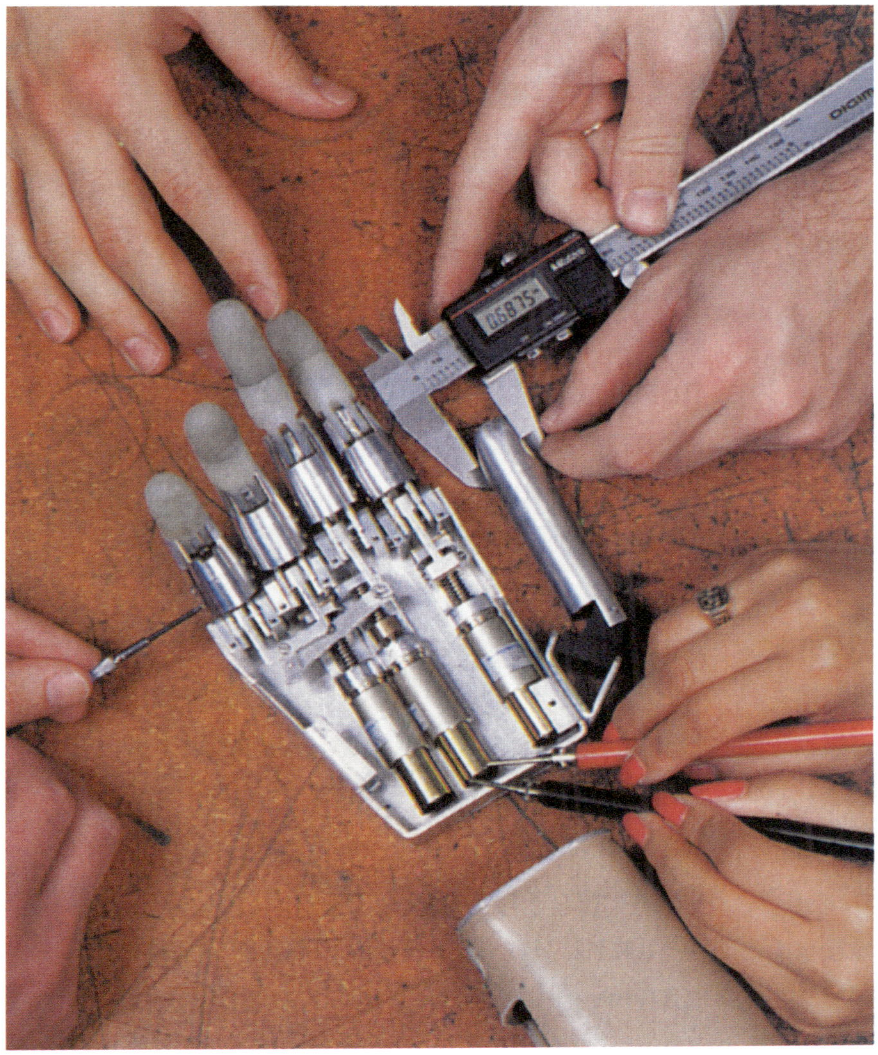

medicine, and synthetic biology, with a particular focus on cancer-related research. And Ensor focused on time series, spatial processes, and stochastic processes—applying statistics to dynamic systems across fields like health, energy, and the environment. Together, their complementary strengths laid the groundwork for a department that would help define data science as both a discipline and a driver of innovation.

## Robot Hands and Nano Plans

Rice researchers in mechanical engineering and materials science embraced these new innovations to develop medical robotics for surgery and prosthetics.

In 1991, a Houston-based foundation called Limbs of Love held a two-day meeting at NASA, bringing together amputees, doctors, engineers, and industry experts to address the high cost and poor quality of prosthetic devices. The need for better solutions called engineers to action—and Rice responded.

By 1993, Rice students had developed a prototype of robotic hands designed to aid individuals who had lost the use of a limb. Under the guidance of professor and Rice alumnus John Cheatham, undergraduate students engineered a system that highlighted the critical role of sensory input in grasping techniques. Their work was supported by faculty across disciplines.

Ian Walker from electrical and computer engineering tackled the mathematics of controlling robotic grasps. Through the Computer and Information Technology Institute, Don

"

His pioneerin
and mentors
a foundation
continues to
engineering

g work
hip built
that
shape
today.

**Don Johnson**

Don Johnson, who joined the Rice faculty in 1977, is widely recognized for his pioneering contributions to digital signal processing and his lasting influence as a researcher, educator, and mentor.

Johnson's research on auditory signal processing laid the foundation for the mathematical and statistical frameworks that explained how neurons in the brain receive, encode, process, and respond to sensory cues. This work is helping to advance the development of neural prosthetics to treat neurological and psychological disorders. His contributions in non-Gaussian statistics have led to significant improvements in functional magnetic resonance imaging (fMRI) used in healthcare and biomedical research.

He coauthored the seminal 1993 textbook *Array Signal Processing: Concepts and Techniques*, the first to synthesize this subject area. A legendary educator, Johnson received Rice's George R. Brown Award for Excellence in Teaching and four George R. Brown Superior Teaching Awards.

> We're trying to make machines more intelligent—to combine mechanical design and computer control so the machine can function as an assistant for people.

**John Cheatham,** professor of mechanical engineering

Johnson and three-time Rice graduate James D. Wise led software development to operate the hand. Collaborators from Baylor College of Medicine, NASA, and the Institute for Rehabilitation and Research added clinical and technical insight.

"We're looking for robotics to give assistance to amputees and quadriplegics," Cheatham said. "What we're trying to do is to make machines more intelligent—to combine mechanical design and computer control so the machine can function as an assistant for people."

During this same decade, the stage was set for Rice to lead in another transformative space— nanotechnology. In his 1992–93 Report of the President, university president Malcolm Gillis observed, "Within the past few years, an entirely new field has emerged, focused upon matter at the molecular and atomic scale. This emerging field is called nanoscience: the study of structures of a size between one and one thousand nanometers, where a nanometer is one billionth of a meter. From nanoscience will come nanotechnology, the ability to engineer matter on a nanometer scale for useful purposes. Nanotechnology is truly interdisciplinary in approach. It is the outgrowth of a synthesis of physics, chemistry, microelectronics, materials science, and biology. Potential applications of successful nanotechnology are staggering, not only in supercomputing but in medicine, optics, and molecular biology."

As computing power increased and component sizes became smaller, demand grew for advanced nanomaterials. Rice answered this demand with the launch of the Center for Nanoscale Science and Technology in 1993, later renamed the Smalley Institute in honor of Richard Smalley. It eventually merged with the Rice Quantum Institute to become the Smalley-Curl Institute in 2015 to properly recognize Robert Curl, one of the founders of the Quantum Institute.

Rice focused on the nano frontier, not only in its facilities but also in faculty recruitment

→ The 1992–93 President's Report by Malcolm Gillis recognized nanoscience as an emerging interdisciplinary field with applications in computing, medicine, and molecular biology.

1993 Report of the President:

# Rededication, Reexamination, and Redefinition

Rice University

# ↗ The Deans List

## Michael M. Carroll
(1988–1997)

During his tenure, Michael M. Carroll significantly expanded the scope and stature of Rice Engineering. He spearheaded the establishment of the Department of Bioengineering in 1997, which has become one of the university's top-ranked programs. The engineering faculty grew from seventy-four to ninety-two members, and the percentage of female undergraduate engineering students increased from 17 to 30 percent. Carroll was known for recruiting outstanding faculty, building strong external support, and bringing recognition to his faculty, including elections to the National Academy of Engineering. A prolific scholar, Carroll authored more than one hundred academic papers in areas ranging from nonlinear optics to the mechanics of sports. He has been a consultant to the Livermore and Sandia National Laboratories and has served on government boards for the National Science Foundation, the National Research Council, and the Advanced Research Projects Agency. He was a member of the National Academy of Engineering and a fellow of the American Academy of Arts and Sciences.

Michael M. Carroll

and the purchase of instruments. The Department of Mechanical Engineering and Materials Science hired its first nanotechnologist, Bruce Lairson, in 1993 to develop ultra-thin films of magnetic material for computer hard disks. About the same time, Enrique Barrera's work on lightweight material for airplane wings uncovered mechanical properties that could be used to develop new types of electronic devices. Barrera also worked on developing lightweight fullerene-reinforced aluminum for NASA astronaut gear.

## Engineering Environmental Solutions

Meanwhile, back in the Department of Environmental Science and Engineering, Herb Ward not only voiced optimism as efforts to fight pollution moved forward, he also delivered solutions. "It used to be said that, once contaminated, groundwater is contaminated forever; but that's not true," Ward said, speaking in 1987. "We have worked on remediation processes, and one of the exciting things is that we are using microbial biotechnology to restore groundwater right where it is."

Philip Bedient was going to school in flood-prone Gainesville, Florida, when he chose the path of environmental engineering in the early 1970s. After receiving his doctorate at the University of Florida, he accepted a position at Rice University.

Bedient discovered that there was a need to develop nature-based solutions for city infrastructure in Texas and at Rice, where he taught and researched surface-water and groundwater hydrology, flood prediction and warning systems, and water-quality assessments. In 1992, he took on one of the biggest challenges ever presented in environmental engineering: solving the problem of pollution by dense non-aqueous phase liquid (DNAPL), a chlorinated solvent then considered the number one contaminant in the United States.

"We saw massive plume contaminations—ten miles long in cities like Phoenix, and widespread across California and the northeast," Bedient said.

↑ Rice alumnus Keith Cooper founded the Compiler Research Group and later served as chair of the computer science department and associate dean for research.

He and his collaborators further burnished Rice's reputation as a leader in groundwater research, urban hydrology, and flood prediction.

## Computing Powers Up

When the Department of Computer Science became a reality in March 1984, mathematical sciences professor Ken Kennedy, who had graduated from Rice in 1967, served as chair. Two years later, Kennedy and Sidney Burrus cofounded the Computer and Information Technology Institute (CITI), with Kennedy serving as director.

"Ken was an incredibly bright man," said Keith Cooper, L. John and Ann H. Doerr Professor in Computational Engineering and professor of computer science. Cooper, a three-time Rice graduate, completed his PhD and postdoctoral research with Kennedy before joining the faculty and becoming Kennedy's colleague and friend. "He had a very quick intelligence. Ken would get running and be a mile ahead of you, and then stop and let you go think, and come back and tell him why he was wrong. He was willing to back off his convictions and change his mind, in the face of evidence. He was a visionary. He believed software could do anything."

Kennedy did an effective job persuading the board at the National Science Foundation with a

In 1985, computer science students work in the lab, with some using the newly introduced Apple computers.

Thm

> CRPC cemented Ken Kennedy's position as a leader in the field of computing … it was Rice Engineering's first real splash in the national and international community.

**Bart Sinclair,** professor emeritus of electrical and computer engineering

← Founder of Rice's computer science department, Professor Ken Kennedy lectures before students at the blackboard.

proposal to build and operate a $5-million science and technology center: the Center for Research on Parallel Computation (CRPC). "Ken's field was parallel processing, and his argument was, 'We want to be able to solve the equations of weather before the weather happens. How can we do that?' " said Don Johnson, who was the document manager for the proposal and went on to have a storied career at Rice. "That's parallel processing. It was a big open topic in programming."

The National Science Foundation ultimately selected Rice as the site for the center. The opening of the CRPC in 1989, one of twenty-five science and technology centers supported by the National Science Foundation, aimed to develop software and operating systems that use multiple processors simultaneously—making massively parallel computing systems as usable as conventional supercomputing systems. With a $4.7-million annual stipend, the CRPC drove the national agenda in parallel computing.

The CRPC's research focused on the development of precompetitive software, algorithms for high-performance computing, and the integration of diverse computing environments. Its Geophysical Parallel Computation Project, for example, directly affected the process of oil recovery.

"CRPC cemented Ken Kennedy's position as a leader in the field of computing," professor emeritus Bart Sinclair said. "It brought a great deal of attention to Rice. From my perspective, it was Rice Engineering's first real splash in the national and international community."

Likewise, Kennedy's contributions elevated Rice's reputation in the computing field. "He led the largest NSF-funded grant in the country and co-chaired Bill Clinton's President's Information Technology Advisory Committee," Cooper said. "He left behind a series of amazing PhD students who've had a strong impact on the field. His technical accomplishments were strong, but it was his ability to influence policy and to change the direction of other researchers in the field that really had greater significance. He believed that computer science would cure all of society's ills."

As Rice invested in computational research, the Department of Mathematical Sciences responded to this shift. In 1991, it was renamed the Department of Computational and Applied Mathematics to reflect its increasing emphasis on numerical methods, scientific computing, and algorithmic problem-solving. The change represented a broader reorientation of applied math as a central player in both engineering and science, particularly in an era increasingly defined by computation.

During this same period, Robert Bixby, a professor in the Department of Computational and Applied Mathematics, was laying the groundwork for transformative tools in mathematical optimization. In 1987, he cofounded CPLEX®, the first commercial optimization solver to incorporate advanced algorithms for linear programming, setting a new standard for performance and scalability. His research and innovations at Rice would later lead to the cofounding of Gurobi®, now one of the world's fastest and most widely used solvers. Together, these innovations have revolutionized industries from logistics to energy—solidifying Rice's role in shaping the future of computational mathematics and real-world problem-solving.

ons

At Rice Engineering and Computing, learning comes alive through action. Students tackle real-world challenges, lead design projects, and dive into hands-on experiences that build lasting connections with classmates. These opportunities do more than teach skills—they spark growth, creativity, and the confidence to make an impact on a global scale. Much of this experiential learning is fueled by the strength of the alumni community. By mentoring students, sharing expertise, and building bridges to industry, Rice Engineering and Computing alumni ensure that the next generation is not only prepared for the future but ready to lead it.

1

2

3

4

1 In the lab of associate professor Jackie Shanks, freshman Rafael Verduzco—now a Rice professor—studies for an Amoco project how plants can clean up TNT.

2 Professor C. Fred Higgs III talks with students in the Mechanical Engineering Building.

3 Members of the Data to Knowledge Lab, which tackles real-world data science challenges, gather for a celebratory photo.

4 Students gather for a study session atop one of the monoliths in the Engineering Quadrangle.

5 Students compete in the Rice Center for Engineering Leadership's Splashdown Design Challenge, attempting to cross the university's recreation center pool without getting wet.

5

The First

Engineering Show

of

The Rice Institute

MAY SEVENTH, 1920

FROM 2 P. M. TO 10 P. M.

American Indemnity Company

MILTON P. IVEY, Manager

Engineers' and Contractors' Surety Bonds

Automobile and Workmen's Compensation Insurance

GENERAL PUBLIC LIABILITY AND BURGLARY INSURANCE

2

3

**1** Rice Eclipse, the university's rocketry team, became the first in Texas to launch a student-developed hybrid engine, winning top honors at the 2024 Spaceport America Cup.

**2** Beginning in 1920, Rice students hosted a biennial Engineering Show, drawing more than ten thousand visitors to its inaugural event.

**3** Students socialize before a networking event.

**4** The Rice Solar Car team, an interdisciplinary group of engineering and science students, works to improve the design of its solar-powered vehicle.

**5** At the annual School of Engineering/Rice Engineering Alumni picnic in 2016, more than $165,000 in scholarships, grants, and awards were distributed. REA president Wendy Hoenig poses with students.

4

5

↗ 1995–2004

# From Research to Real-World Impact

**As Rice Engineering** entered the new millennium, three powerful frontiers—bioengineering, nanotechnology, and environmental systems—defined its trajectory. These fields shaped research, education, and institutional growth, positioning the school as a leader in solving twenty-first-century global challenges.

Malcolm Gillis, president of Rice University from 1993 to 2004, believed research was the path to progress. In a 2003 paper for the Federal Reserve Bank of Dallas titled "Harnessing New Technologies for the 21st Century," he noted, "New science-based industries will have to play roles comparable with the chemical industry after 1900." He championed Rice's rising reputation as a research hub.

Breakthroughs, discoveries, and solutions to once-insurmountable problems became hallmarks of the school's work. New departments formed. Major grants were secured. Interdisciplinary research blossomed. And the physical campus transformed to match the school's ambition.

## Built to Heal

The Department of Bioengineering, established in 1997, began with just twenty undergraduate students and a small founding faculty. Its launch came on the heels of a $5-million NASA grant awarded to Rice's Institute for Biosciences and Bioengineering to create a NASA Specialized Center of Research and Training, or NSCORT, in gravitational biology. The center promoted collaborations between Rice engineers and scientists and their colleagues at the nearby Texas Medical Center, the Johnson Space Center, and in private industry.

→ Rice engineers are developing new ways to reclaim and purify previously discarded water, including municipal and industrial wastewater, brackish groundwater, and seawater.

↑ An interior view of Anne and Charles Duncan Hall highlights the ceiling mural, *The Birth of Consciousness.*

Research in the department quickly made its mark. The study of bone formation led to a pioneering method for bone regeneration, achieved by seeding bone cells onto an injectable, biodegradable polymer scaffold. Projects on joint repair and nanotechnology addressed osteoarthritis by developing methods to grow replacement cartilage for the meniscus, the wedge of cartilage that cushions the knee. Studies into the effects of gravity on immune cells proved invaluable to NASA. These efforts reflected a broader trend—engineers working closely with biologists, chemists, and medical researchers.

In 2004, Rice officially became a member institution of the Texas Medical Center, formalizing the relationship with the world's largest medical complex that dated back to the 1960s. Rice's location at the doorstep of the medical center made it possible to take technology developed in the lab out into the world.

As the School of Engineering grew, so did its need for space. Named for Charles Duncan, a 1947 graduate of Rice and former chair of the Rice University Board of Governors, and his wife, Anne, Duncan Hall was dedicated in 1996, providing a new home for the departments of computer science, computational and applied mathematics, statistics, and part of the electrical and computer engineering department.

Made possible through the generosity of many donors, including Josephine Abercrombie, John L. Cox, Charles Duncan, Houston Endowment, the Keck Foundation, Arco, and Texas Instruments, Duncan Hall was designed to inspire collaboration. The gift supported PhD researchers, visiting professorships, and joint work in digital signal processing and information engineering. Architect John Outram's design emphasized flow and openness. The space encouraged spontaneous interaction, with offices grouped by

## To take a complex idea and make it accessible, that's the challenge.

**Sidney Burrus,** dean of engineering

↓ Professor Sidney Burrus, a Rice alum, joined the Rice University faculty in 1965 and taught electrical engineering.

↓↓ NASA leadership meets with Rice faculty, including professors Sidney Burrus (center) and Richard Tapia (far right), to discuss a potential collaboration to develop aerospace applications of carbon nanotubes.

research area and shared resources like coffee machines placed strategically to promote casual conversation.

Hovering over the atrium, a ceiling mural titled *The Birth of Consciousness* was printed on vinyl panels and wrapped around acoustical tiles—an impressive engineering design executed for less than the cost of the building's door handles.

### The Burrus Effect

As the school expanded, so did its leadership. In 1998, Sidney Burrus—a Rice graduate and longtime faculty member—was named dean of the George R. Brown School of Engineering. He had created the first course in digital signal processing (DSP) at Rice in 1968, and his research with Tom Parks helped establish Rice's leadership in the field.

"Part of my own reason for being an academic is that I like to make complicated ideas accessible to practicing engineers and to students who are not as abstract as some faculty members are," Burrus said. "To take a complex idea and make it accessible, that's the challenge [for] engineering faculty and . . . researchers."

One of Burrus's legacies during his tenure was renewing the focus on chemical engineering. The Department of Chemical Engineering was renamed the Department of Chemical and

"

From complex
challenges to
solutions, they
advance engi
on a global sc

1

2

practical

eering

ale.

**Pedro J. J. Alvarez** [1]

Pedro J. J. Alvarez, George R. Brown Professor of Civil and Environmental Engineering at Rice, is a global leader in sustainable water treatment. Alvarez serves as the founding director of the NSF Engineering Research Center for Nanotechnology-Enabled Water Treatment (NEWT), a national hub for clean water innovation. He is also the director of the Water Technologies Entrepreneurship and Research (WaTER) Institute, which is dedicated to advancing research and innovation to shape the future of sustainable water treatment across three key focus areas—public health, energy transition, and resilient infrastructure.

**Naomi Halas** [2]

An engineer, chemist, physicist, and pioneer in the field of nanophotonics, Naomi Halas is a former director of the Smalley-Curl Institute, one of Rice's largest interdisciplinary institutes. Her research transcends the boundaries of schools and disciplines, providing new insights into how light and matter interact at the smallest scales, driving advances in cancer therapy, chemical sensing, catalysis, and water treatment. Halas has cofounded two companies: Nanospectra Biosciences, which focuses on localized cancer therapeutics, and Syzygy Plasmonics, which develops low-cost, light-driven hydrogen production. She was elected to the National Academy of Sciences and to the National Academy of Engineering. She is also a member of the American Academy of Arts and Sciences and the National Academy of Inventors.

> The goal of Connexions was to break the bottleneck associated with conventional, paper-based textbooks.
>
> **Richard Baraniuk,** C. Sidney Burrus Professor of Electrical and Computer Engineering

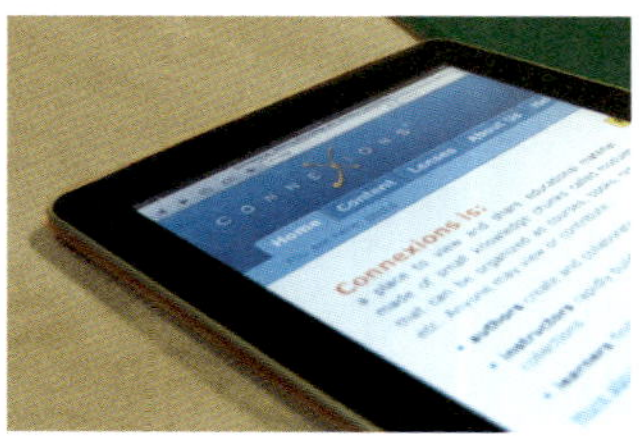

↑↑ Details about the Connexions system are displayed on a tablet.

↑ Professor Richard Baraniuk founded Connexions, an adaptive e-textbook system transforming education with personalized, open-source learning.

Biomolecular Engineering, reinforcing the belief that integrating biological sciences with chemical engineering would lead to groundbreaking discoveries. The department's work in these areas strengthened its reputation nationally and internationally.

As the department's visibility grew, so did student success. In 1998, Rice Engineering undergraduates were awarded twenty-seven National Science Foundation fellowships—the most in the university's history—placing Rice first in the nation in the percentage of students who received this honor.

Burrus's research bona fides were as impeccable as his teaching and organizational skills. When Texas Instruments created the Leadership University Program to "lead TI into the future of DSP," the idea was to give a professor at a chosen university $1 million over three years, with no specific research goals, no strings attached. Rice received the award multiple times. In each instance, the university's faculty awardee was Sidney Burrus.

## Open Knowledge, Open Access

Another outgrowth of the school's collaborative and future-focused culture was a bold effort to rethink how students access educational resources. In 1999, Richard Baraniuk, C. Sidney Burrus Professor of Electrical and Computer Engineering, recognized a growing challenge: the rising cost of textbooks was placing an unsustainable burden on students around the world. With encouragement from Rice Engineering leaders and alumni—including Sidney Burrus, Don Johnson, and advisors Robert Maxfield '63 and Bill Sick '57—Baraniuk launched Connexions, a free digital repository of educational content that predated Wikipedia and MIT OpenCourseWare. Connexions ultimately produced more than twenty-four thousand learning modules in forty languages and helped lay the foundation for the global open educational resources (OER) movement.

That early vision evolved into OpenStax, founded in 2012 as a nonprofit educational initiative at Rice. Today, OpenStax is the world's largest publisher of free OER, a provider of digital learning technologies, and a leader in education research. Its eighty titles—covering high school and college-level material—are widely used by students and educators in the United States and abroad. Since its inception, OpenStax has served more than thirty-six million students, saving them $2.9 billion in educational materials. Its resources are now used at two thousand institutions in 153 countries, including by four million students outside the United States.

OpenStax's mission goes far beyond access. Under Baraniuk's leadership, the initiative is leveraging AI-supported tools to deliver personalized, culturally relevant course materials proven to be more effective for student learning. With a recent $90-million grant from the National Science Foundation—the largest investment in educational research infrastructure in the United States—OpenStax is leading a multi-institutional effort to build SafeInsights, a large-scale, secure digital platform for learning and education research. It is poised to transform how students and educators experience education.

"We are focused on removing barriers to education and cultivating a generation of learners who are engaged, resilient, and prepared to create positive change," Baraniuk said.

# ↗ The Deans List

## C. Sidney Burrus
(1998–2004)

C. Sidney Burrus was an innovator in digital signal processing (DSP) whose scholarship and teaching helped shape Rice's engineering legacy. In the 1980s, he coauthored two landmark books that advanced a unified theory of fast Fourier transforms (FFTs) and developed more efficient ways to program them—solidifying Rice's standing as an authority in the field. He created the first DSP course at Rice in 1968 and, over his fifty-six-year career on the faculty, won the George R. Brown Teaching Award six times.

Burrus held multiple roles, including chair of electrical and computer engineering, director of the Computer and Information Technology Institute, and dean of engineering. A fellow of the American Association for the Advancement of Science, he received the 2009 Jack S. Kilby Signal Processing Medal from the Institute of Electrical and Electronics Engineers for his contributions to FFT algorithms, digital filter design, and signal processing education. In 2020, Ann and John Doerr honored his legacy by establishing the Sidney Burrus Chair in electrical and computer engineering.

**C. Sidney Burrus**

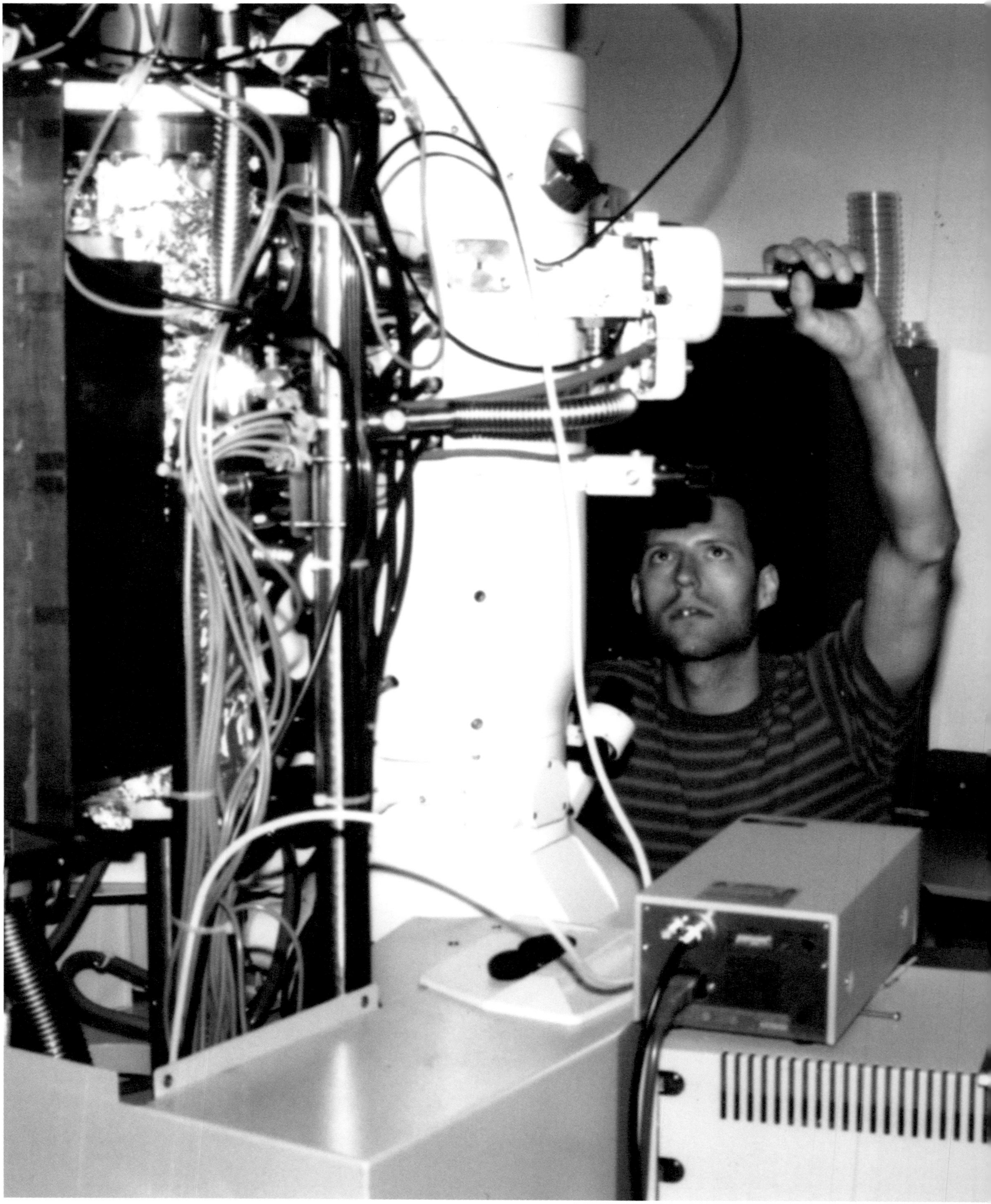

"Ultimately, our goal is to equip students with the skills and knowledge they need to thrive and shape a more equitable society."

## Nanoscale, Macro Change

Just as Rice engineers helped shape the field of digital signal processing, they accelerated the university's growing influence in the evolving field of nanotechnology.

The Center for Nanoscale Science and Technology (CNST) was the first center for nanoscale inquiry in the world. Founded by Rice chemist Richard Smalley in 1993, CNST fostered interdisciplinary research and opened the door for Rice engineers to use nanotechnology to solve challenges in medicine, materials, and energy.

Wade Adams, a professor of chemistry and former senior executive in the Air Force Research Laboratory at Wright-Patterson Air Force Base, succeeded Smalley as director of CNST in 2002. A decade later, he became associate dean of Rice's School of Engineering.

Naomi Halas, a professor in electrical and computer engineering, and bioengineering professor Jennifer West furthered applications of nanomaterials in the medical field, using light-absorbing nanoshells to target and eliminate cancer cells with infrared light. The pair also developed a whole-blood immunoassay with nanoshells that could be used to diagnose conditions ranging from HIV to heart attacks.

## Resilient by Design

Environmental challenges were growing more urgent—and engineering solutions more essential. Persistent flooding in Houston, and around the Texas Medical Center in particular, prompted professor Philip Bedient to develop a real-time flood warning system in 1999 based on radar rainfall. The system proved its value when Tropical Storm Allison brought significant rain to Houston, and it was later embraced by the Texas Department of Transportation.

In 2001, Rice merged its civil engineering department with its environmental science and engineering department. The new Department of Civil and Environmental Engineering broadened its scope and strengthened its

↑ Professor George Hirasaki and doctoral student Robert Szafranski (from left to right) tested new technology to clean up toxic substances as part of a Department of Defense project to eliminate hazardous waste on military bases.

← A graduate student operates a transmission electron microscope.

> We ended up creating a whole new approach to training civil and environmental engineers.

**Philip Bedient,** Herman Brown Professor of Engineering

← Professor Philip Bedient visits Brays Bayou in 1997.

↓ Civil engineering students work on a project in the basement of the Ryon Engineering Laboratory.

contributions to sustainable urban systems and resilient infrastructure.

"We ended up creating a whole new approach to training civil and environmental engineers," Bedient said. "Because even though they might prefer one or the other, they would invariably go out into the working world and have to address both. It was a marked change in our educational initiatives for the department."

The newly combined department was home to the National Center for Groundwater Research, a field of great interest to Pedro J. J. Alvarez, who joined the faculty in 2004 and became the department's chair in 2005.

### Collaboration Is the Culture

Burrus wasn't alone in seeing the power of collaboration. "The collaboration that existed between faculty across department lines within engineering was important, but also those collaborations that crossed school lines—from natural sciences to humanities and social sciences," Bart Sinclair explained. "After I had been associate dean and met with other associate deans and financial managers in top-tier engineering schools, I realized that the atmosphere at Rice was not typical."

Pedro Alvarez summed it up well, saying, "We trust each other. We lower our guard to share information and collaborate."

Rice Engineering's collaborative culture and ability to anticipate the next wave—whether digital, biological, or environmental—positioned it as a powerhouse for twenty-first-century innovation.

# Age of Computing

→ Jane Griffin works at the control terminal of the R1 computer as Martin Graham observes. Designed and built by Rice faculty, the R1 was the university's first large computational research computer.

When the R1 computer arrived at Rice in 1956, it opened the door to a new era of computational research. Over the following decades, Rice engineers and researchers excelled in digital signal processing, computer systems, artificial intelligence, programming languages, and computational optimization, shaping the school's reputation as a leader in computing. In 2024, the renaming to the George R. Brown School of Engineering and Computing celebrated fifty years of excellence in the field. Today, Rice researchers harness computing to tackle challenges in public health, disaster recovery, and energy and sustainability, remaining at the forefront of emerging technologies.

1

2

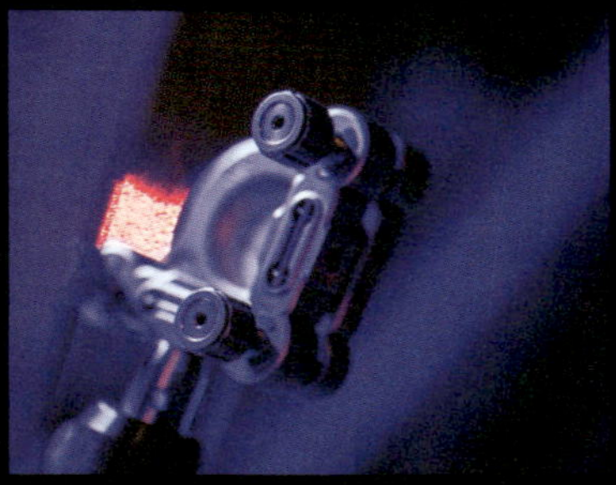

3

**1** A new Rice-developed machine learning algorithm interprets optical spectra with exceptional accuracy, opening doors to faster medical diagnoses and more precise sample analysis.

**2** Research from Professor Shengxi Huang and collaborators indicates that carbon-doped hexagonal boron nitride shows promise as a quantum tech platform.

**3** Professors Peter Norlander and Moshe Vardi and research scientist Robert Fowler (left to right) stand with the IBM supercomputer Rice acquired in 1998 to be used by researchers across campus.

**4** Rice engineers have reimagined the digital camera, replacing power-hungry microprocessors with new mathematics and a silicon chip of microscopic mirrors that reconstruct high-resolution images from thousands of single-pixel snapshots—dramatically boosting efficiency.

**5** Professor Joseph Cavallaro (left) demonstrates a new Texas Instruments digital signal processing board as Gene Frantz, TI fellow, and Patrick Frantz, executive director of Rice's Center for Multimedia Communications (both seated), review programs written by students in Rice's digital signal processing course.

**6** Ashutosh Sabharwal and colleagues created WARP, an open-source platform that enables researchers to test high-speed wireless technologies without building test systems from scratch.

**7** Don Johnson and William Wilson, known affectionately to students as "Dr. Bill," share a laugh in front of a Visual 200 computer.

4

5

6

7

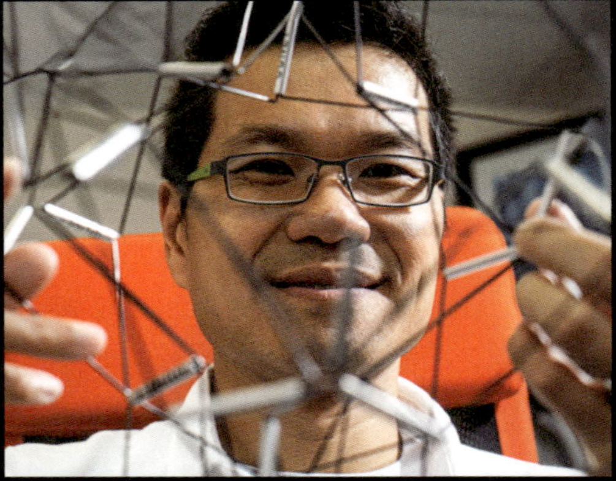

1

2

3

**1** Professor Eugene Ng and his research team develop strategies that enable networks to dynamically adapt their structures, which accelerates transmissions of large data sets.

**2** William "Dr. Bill" Wilson works with Greg Shinn in a semiconductor lab.

**3** A computer engineer uses a Visual 200—an early computer with a cathode-ray tube display and no graphics capabilities.

**4** Professor Larry McIntire chaired both the Chemical Engineering and Bioengineering departments and served as director of Rice's Biomedical Engineering Laboratory, where computing played an increasing role in research.

**5** Research from Songtao Chen and his colleagues shows promise for advancing quantum networks with a T-center-based silicon quantum photonic chip.

4

5

2005–2014

# Hands-On Research, Practical Solutions

**The arrival of** Sallie Keller, a statistician, as dean of engineering in 2005 marked another period of growth for the George R. Brown School of Engineering. At the time she was hired, Rice was one of only four American universities that included statistics within engineering. When announcing her as the new dean, Rice President David W. Leebron acknowledged that she was an out-of-the-box choice: "However, engineering needs out-of-the-box thinkers, and Rice and its engineering school have them."

To support Keller's vision and leadership, electrical engineering alumnus William "Bill" Sick '57 and his wife, Stephanie, increased their commitment to the William and Stephanie Sick Professorship to make it the permanent chair held by the dean of engineering. Keller became the first dean of engineering to hold the chair.

Dean Keller inherited and championed a culture of interdisciplinary collaboration that had already begun taking root at Rice. One early example was the Center for Biological and Environmental Nanotechnology, founded in 2001 by Kristen Kulinowski, a chemist in the School of Natural Sciences. Though not an engineer, Kulinowski embodied the collaborative spirit that would define the coming decade—bringing together scientists and engineers to tackle real-world challenges at the molecular level.

Five years later, in 2006, Rice chemists and engineers won a $12-million renewal grant from the National Science Foundation, turning the theoretical goal of creating sustainable nanotechnologies that improve human health and the environment into an achievable reality.

The funding provided a platform and the impetus for researchers from engineering,

→ A student researcher explores biomedical solutions in a lab at Rice's BioScience Research Collaborative.

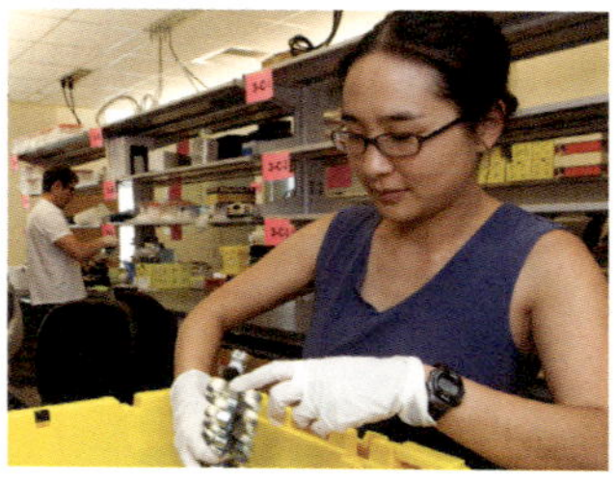

↑ Faculty and students move into the BioScience Research Collaborative at the Texas Medical Center.

↑↑ The BioScience Research Collaborative provides ten floors of laboratories and centers for interdisciplinary research and innovation.

biological sciences, and information technology to engage in interdisciplinary systems-level research, while working to develop sustainable nanomanufacturing systems and to provide incentives for the academic entrepreneur. Through the development of biomedical technologies, nanotechnology was revolutionizing how diseases could be detected and cured, and how water and waste streams could be more effectively cleaned and remediated.

Beyond medical and environmental applications, nanoscale engineering was also being used in everyday products like clothing, furniture, and household goods.

For these reasons, the Center for Biological and Environmental Nanotechnology devoted considerable resources to educating teachers, students, and the public to the possibilities of nanotechnology. If more people understood the potential of nanotechnology, then that potential would be limited only by the imagination.

## From Lab to Life

Another interdisciplinary institute, the Rice360 Institute for Global Health Technologies, was launched in 2007 as part of a $100-million global health initiative by Rice University. The hub for science, engineering, public policy, social sciences, and humanities was formed to combat child mortality and various critical health issues for women and children in the developing world.

While Rice360 focused on low-cost technologies for underserved communities abroad, Rice also expanded its presence locally through biomedical innovation hubs.

In 2009, Rice University opened the BioScience Research Collaborative (BRC) at the Texas Medical Center (TMC) as a home for bioengineering research partnerships with other TMC institutions. The BRC would become a magnet for the "bench to bedside" perspective, bringing projects from research to application with real medical relevance and making Texas a pipeline for new medical technology.

# ↗ The Deans List

## Sallie Keller
(2005–2010)

As dean, Sallie Keller championed experiential learning and interdisciplinary collaboration. She secured funding and space for the Oshman Engineering Design Kitchen, emphasizing the role of design across all eight engineering departments. She also laid the foundation for the Rice Center for Engineering Leadership, supported by gifts from alumni Ann '75 and John '73 Doerr. Under her leadership, research expenditures exceeded $50 million and doctoral enrollment rose by 20 percent. Keller supported collaborative hires with Texas Children's Hospital and Baylor College of Medicine, expanding Rice's role in neurological research. She is a member of the National Academy of Engineering.

**Sallie Keller**

## Edwin L. Thomas
(2011–2016)

During his tenure, Edwin L. Thomas strengthened faculty excellence and oversaw the creation of the Department of Materials Science and NanoEngineering after its separation from mechanical engineering. He also established the Engineering Advisory Board and supported the launch of the center for Nanotechnology-Enabled Water Treatment (NEWT) and an interdisciplinary doctoral program in systems, synthetics and physical biology. Graduate degrees awarded to women rose above 30 percent, higher than the national average of 23 percent. A member of the National Academy of Engineering, Thomas oversaw the expansion of the Oshman Engineering Design Kitchen and the development of the Rice Center for Engineering Leadership's certificate program.

**Edwin L. Thomas**

"

Their work exp
what engineer
can achieve, le
a lasting impa
technology, so
and the profes

ands
ng
aving
ct on
ciety,
sion.

1

2

**Lydia Kavraki [1]**

Lydia Kavraki, Kenneth and Audrey Kennedy Professor of Computing, and director of the Ken Kennedy Institute, is internationally recognized for her research breakthroughs in robotics, artificial intelligence and computational biomedicine. She has been elected to the American Academy of Arts and Sciences and is the first Rice faculty member elected to all three US National Academies. In robotics, she develops advanced methodologies to enhance human-robot interaction and collaboration. In computational biomedicine, she applies AI-driven approaches to model biomolecular interactions and supports the design of novel therapeutics.

**Rebecca Richards-Kortum [2]**

Rebecca Richards-Kortum, Malcolm Gillis University Professor and codirector of the Rice360 Institute for Global Health Technologies, is a leader in developing low-cost medical technologies for underserved communities. Her team's innovations, designed for low-resource settings, include a battery-powered system to detect precancerous tissue without biopsies, neonatal breathing devices, BiliSpec for jaundice detection, and the DoseRight Syringe Clip for pediatric dosing. She has been elected to the National Academy of Engineering, the National Academy of Sciences, and the American Academy of Arts and Sciences. In 2016, she received a MacArthur "genius grant."

An engineering student works on a capstone project using the resources from the Oshman Engineering Design Kitchen.

It takes great leaders (and teams) to change the world. They're the only thing that ever has.

**John Doerr,** Rice electrical engineering alumnus '73 '74

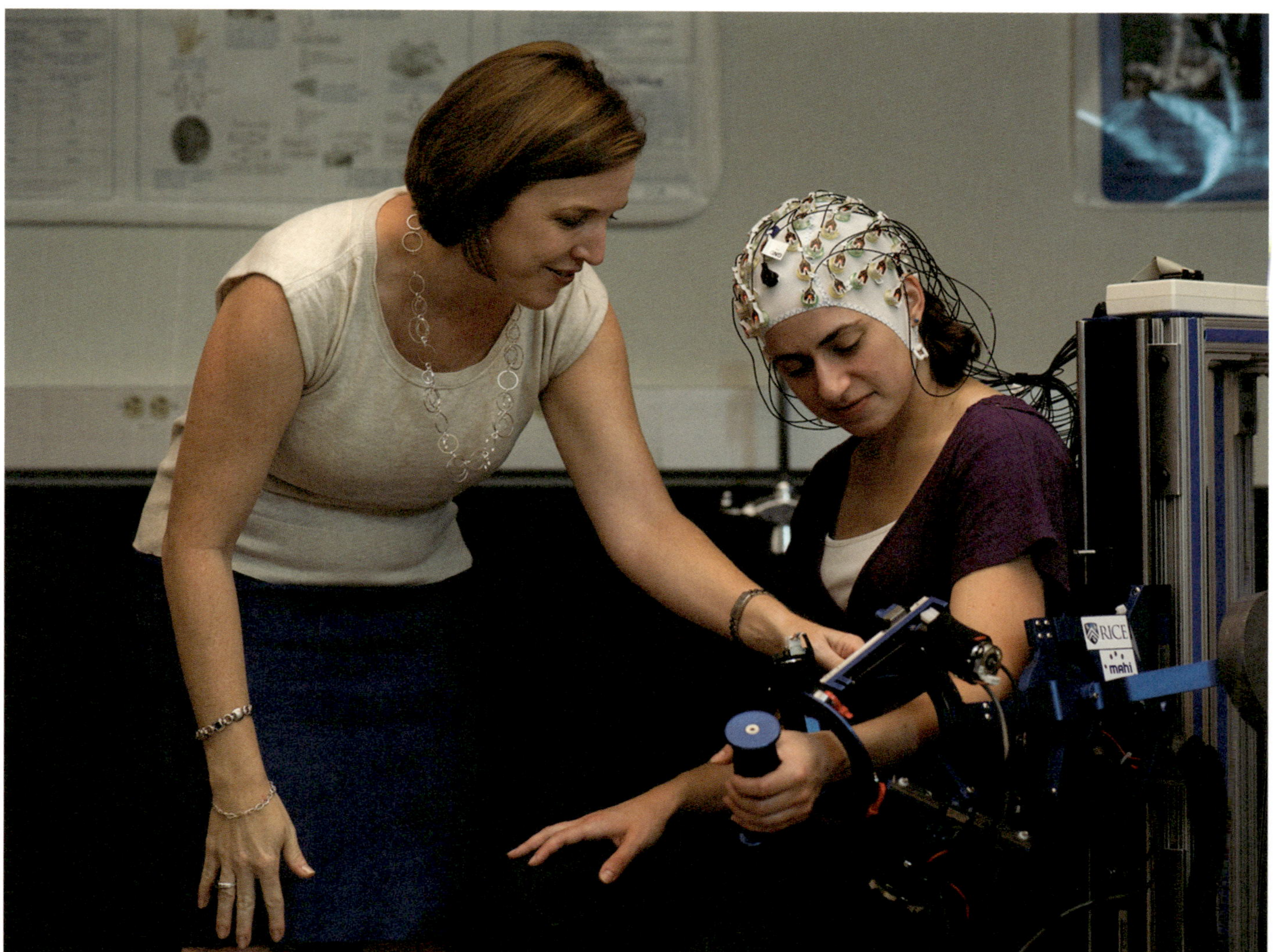

Another significant milestone in health was the success of the EZ-IO device, co-invented by professor Kyriacos Athanasiou. By 2008, the device—which enables emergency medical personnel to inject fluids directly into a patient's bone marrow when intravenous access is difficult—had been adopted by approximately 40 percent of US emergency medical services and by the US armed forces in Iraq and Afghanistan. That year, it received the *Wall Street Journal's* Gold Technology Innovation of the Year award.

## The Making of an Engineer

Dean Keller encouraged the expansion of hands-on opportunities for Rice Engineering students to do real-world research, and she oversaw the development of several experiential learning programs.

In 2008, the Department of Mechanical Engineering transformed Hicks Kitchen, the university's central food-service kitchen, into the Oshman Engineering Design Kitchen (OEDK). The cost of the renovation was offset by a generous lead donation of $2.5 million from Rice Engineering alumnus and ROLM Corporation cofounder Kenneth Oshman '62. The OEDK gave undergraduate engineering students a dedicated space for hands-on research to design, prototype, and deploy solutions to real-world engineering challenges.

"The Oshman Engineering Design Kitchen became a real model for schools across the country who wanted to move their design programs forward in a very similar way," Bart Sinclair said. "A lot of schools saw it as a model for how engineering design really could be enhanced by moving towards a more interdisciplinary, collaborative design process."

That same year, Rice also expanded its efforts to develop student leadership skills alongside technical acumen. Rice electrical engineering alumni John and Ann Doerr, '73 '74 and '75, respectively, made a landmark donation of $15 million to found the Rice Center for Engineering Leadership (RCEL). The center was established to prepare students for life after graduation by teaching leadership and other skills outside of the core engineering curriculum, and to keep Rice graduates at the forefront of new developments in engineering. At the time, Ann noted that the "increasingly complex and global world demands great, ethical leaders," and John said, "It takes great leaders (and teams) to change the world. They're the only thing that ever has."

RCEL introduced Texas's first Certificate in Engineering Leadership program in 2014. Every summer, the center hosts the Rice ELITE Tech Camp summer programs for engineering-focused high school students. These intensive one-week STEM-focused courses demonstrate how the engineering design process can solve practical problems.

## Orchestrating an Expansion

Building on Keller's momentum, Edwin L. Thomas, a materials scientist and mechanical engineer who had been head of the Department of Materials Science and Engineering at the Massachusetts Institute of Technology, came to Rice in 2011 to succeed her as the William and Stephanie Sick Dean of Engineering. Fondly referred to as "Ned," Thomas oversaw the division of the Department of Mechanical Engineering and Materials Science at Rice in 2013 into two separate entities, the Department of Mechanical Engineering and the Department of Materials Science and NanoEngineering.

The creation of two separate departments would expand their respective research potentials in the areas of nanotechnology and robotics. Kathy Ensor, who chaired the Department of Statistics from 1999 to 2013, recalled that "Ned had a really deep understanding of the role of engineering in society, and this came through in all of his decisions and actions."

Thomas believed that, if you wanted to solve the world's problems, you needed engineers. He enthusiastically conveyed this idea to students and faculty alike. In his eyes, being an engineer was not just okay—it was cool. And so was Rice Engineering. His enthusiasm helped define a new era of pride and possibility for the school.

↙ Mechanical engineering professor Marcia O'Malley demonstrates neurotechnology that translates brain waves into arm movement.

**1** Bioengineering students in the Global Medical Innovation Program gather on the patio of the Rice Global Paris Center.

**2** Civil and environmental engineering students and architecture students built a cooling tower prototype at the Rice Global Paris Center.

**3** Rice Engineering and Computing students enjoy the view of the Eiffel Tower.

VEDETTES-DE-PARIS
VEDETTES-DE-PARIS

2

1 Global Medical Innovation Program students explore the Arenal Volcano during study abroad in Costa Rica.

2 Rice President Reginald DesRoches and university leaders visit the Consul General of India, Houston.

3-4 Rice's Engineers Without Borders chapter earned two national awards for bringing clean water to Lucidia Mantilla, Nicaragua, installing a storage tank, piping, a pumping unit, and taps for 450 residents.

5 Students in Rice University's Global Medical Innovation Program enjoy hiking the trails of Hacienda La Chimba coffee plantation in Costa Rica.

6 In 2009, Rice Engineers Without Borders completed a four-year water distribution project in the three-hundred-person village of El Pital, El Salvador. Team members worked side by side with community residents to complete the project.

3

4

5

6

1

2

1 The Rice360 team designs medical devices for low-resource settings, delivering lifesaving technologies that improve healthcare in Africa.

2 A Mehta Rice Engineering Scholar, Atanu Purkayastha researches applications for tin perovskite solar cells in Professor Aditya Mohite's lab.

3 Doctoral student Sikandar Bind, a member of the second Mehta Rice Engineering Scholars cohort, studies the characterization of polyacrylamide for enhanced oil recovery in sandstone reservoirs.

4 Mehta Rice Engineering Scholar Srilekshmi Muraleedharan works to develop an n-type semiconductor from diamond.

5 Maria Oden and Rebecca-Richards Kortum celebrate the opening of the new neonatal ward at Queen Elizabeth Central Hospital in Blantyre, Malawi, in 2016.

3

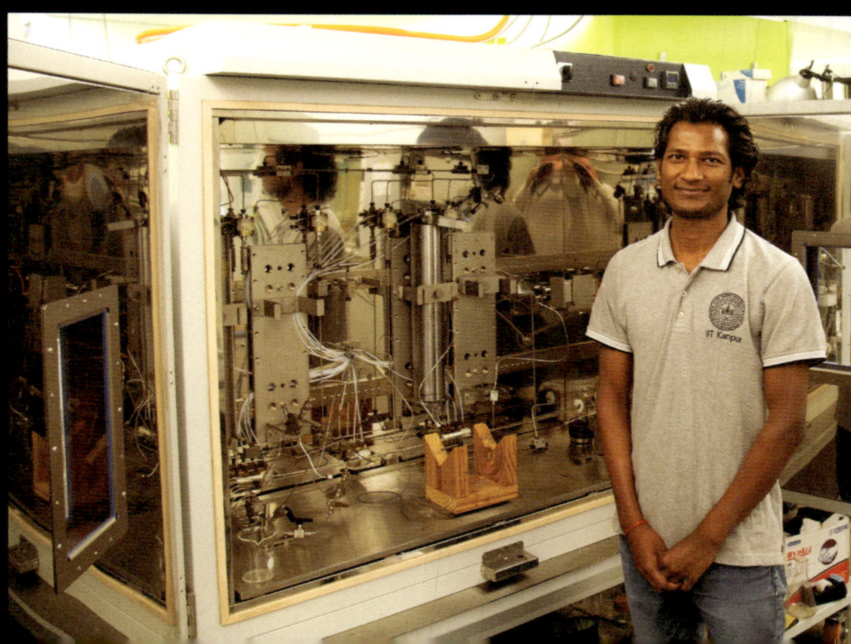

4

5

NURSERY AND KANGAROO MOTHER CARE
BUILT WITH JOINT FUNDING FROM
RICE UNIVERSITY
AND
ELMA PHILANTHROPIES
THROUGH FRIENDS OF SICK CHILDREN
OFFICIALLY OPENED BY
THE PROVOST OF RICE UNIVERSITY
MARIE LYNN MIRANDA
ON THE 29TH OF JUNE 2016

# 15

↗ 2015–2025

# Innovating into the Next Fifty Years

**Rice Engineering has grown** in size, scope, and vision since the school's formation in 1975. Each decade has presented new challenges and new goals, putting the capabilities of Rice engineers to the test. The exponential expansion of the school and its ability to deliver on bold ideas has unfolded at an accelerated pace. Its centers and institutes routinely bring together researchers across disciplines to solve complex problems.

In the decade between 2015 and 2025, the George R. Brown School of Engineering redefined its purpose around a powerful idea: solving for greater good. Inspired by the National Academy of Engineering's Grand Challenges, the school aligned its research and academic mission with some of the most urgent issues facing humanity—from climate change and global health inequities to infrastructure resilience and clean energy. This shift sparked new interdisciplinary collaborations, expanded global partnerships, and deepened the school's commitment to meaningful societal impact.

Today, Rice Engineering is a recognized leader in five research focus areas—health and well-being, energy and sustainability, resilient and adaptive communities, advanced materials, and future computing. These priorities reflect both the school's strategic direction and its enduring principle of solving for greater good.

## Talent and Tools Collide

A reorganization of academic departments created new momentum. The newly independent Department of Mechanical Engineering sharpened its focus on robotics, biomechanics, energy, and aerospace, taking advantage

→ Rice bioengineer Kevin McHugh's lab developed hollow microparticles for timed-release drug delivery. Small enough for hypodermic needles, the dye-loaded particles biodegrade after injection, releasing medication gradually over weeks or months.

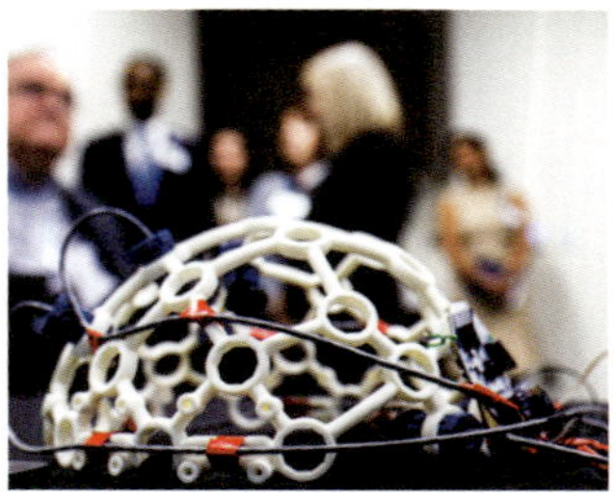

↑↑ At the Center for Neural Systems Restoration, technology aimed at improving brain and spinal cord recovery is displayed, demonstrating the promise of engineering-driven solutions in healthcare.

↑ Tracy Volz guides Rice engineering students in the Activate Communications Program, sharpening their interview skills and empowering them to confidently convey the impact of their work.

→ Salma Alhashim (left) and Sohini Bhattacharyya (right) partner in the Nanomaterials Laboratory led by Rice Professor Pulickel Ajayan to find a faster, cleaner way to extract lithium from battery waste.

of Houston's industry and resources to attract top talent. The university–city dynamic that had helped define Rice Institute from the very start also expanded opportunities for experiential learning and partnerships that shaped curriculum and research.

Advances in instrumentation further accelerated discovery. The arrival of the Titan Themis scanning and transmission electron microscope in 2015 allowed Rice researchers and their academic and industry partners to analyze materials at an astonishing level of detail, down to less than a nanometer. This cutting-edge technology enabled scientists to reliably determine the composition of individual atoms, marking a breakthrough in materials characterization.

Rice recognized that communication is one of the most powerful tools an engineer can use. Launched in 2019, the Activate Engineering Communications Program has helped more than seven hundred students share their research and innovations through courses, workshops, and individualized coaching. The program has grown to include twenty-seven courses and more than twenty-two faculty partners, embedding communication into the heart of engineering education.

## A New Chapter in Leadership

Building on this momentum, a leadership transition brought renewed vision. In 2017, Reginald DesRoches succeeded Edwin Thomas as the William and Stephanie Sick Dean of Engineering, taking charge of nine departments, 137 faculty, and 2,500 students. Under his leadership, Rice Engineering strengthened its commitment to interdisciplinary research—particularly in areas where computing, health, and biology intersect.

## Neuro Joins Engineering

The Rice Neuroengineering Initiative (NEI), established in 2014, exemplified Rice Engineering's interdisciplinary model, bringing together researchers from Rice's engineering departments, the University of Texas Health Science Center at Houston, and Baylor College of Medicine. By focusing on the central and peripheral nervous systems, these researchers laid the groundwork for technologies aimed at diagnosing and treating neural diseases. Early efforts explored noninvasive brain–machine interfaces and the study of neural function to better understand conditions like Alzheimer's, epilepsy, and depression.

"The heart of our mission is very simple. It's to learn how the human brain works," director Benhaam Aazhang asserted.

Among NEI's most ambitious projects is the MOANA (magnetic, optical, and acoustic neural access) headset, a noninvasive device capable of decoding and encoding neural activity using light and magnetic fields, all within less than fifty milliseconds. This technology aims to enable direct brain-to-brain communication without surgery, with the potential to restore sensory functions such as vision.

By 2024, NEI had expanded to include sixteen faculty across departments and occupy more than twenty-five thousand square feet in the BioScience Research Collaborative.

## Expanding the Future of Health

In parallel with NEI's advances, Rice researchers have continued to lead major federal health innovation efforts. In 2024, bioengineer Omid Veiseh secured a $45-million award from the Advanced Research Projects Agency for Health (ARPA-H) to develop a targeted immunotherapy platform for treating tough-to-cure cancers. The grant supported the launch of the Biotech Launch Pad, Rice's first clinical-grade manufacturing facility for developing and testing bioengineered cell therapies.

A Rice-led multi-institutional collaboration also drew national attention in 2024 by receiving $18 million over five years from ARPA-H as part of the federal government's Cancer Moonshot initiative. Its goal is to develop and validate a new system to improve tumor removal accuracy for breast cancer and head and neck cancer.

That same year, Rice and Houston Methodist launched two joint initiatives aimed at

> The heart of our mission is very simple. It's to learn how the human brain works.

**Benhaam Aazhang,** director of the Rice Neuroengineering Initiative

transforming healthcare through interdisciplinary collaboration. The Center for Neural Systems Restoration, housed at Houston Methodist, focuses on advancing treatment and recovery for neurological conditions such as stroke and spinal cord injury. The institutions also established the Digital Health Institute, combining Houston Methodist's strengths in academic medicine and clinical research with Rice's expertise in engineering, digital health, and artificial intelligence to drive innovation across the healthcare spectrum.

Rice University celebrated the opening of the Ralph S. O'Connor Building for Engineering and Science, honoring the late trustee whose generosity supported 250,000 square feet of research labs, classrooms, and collaborative spaces.

↑ Rice University researchers and collaborators have developed a new coating for glass that could help reduce energy bills by preventing heat loss from leaky windows.

↓ Professor Daniel Preston and doctoral student Sofia Urbina design soft machines with programmed failure, enabling smarter systems that can recover from pressure surges and perform multiple tasks using a single control input.

School leadership continued evolving alongside its growth. In 2021, Luay Nakhleh—then chair of the Department of Computer Science—succeeded Reginald DesRoches as the William and Stephanie Sick Dean of Engineering after DesRoches was appointed provost, the chief academic officer at Rice. A computer scientist and computational biologist, Nakhleh was reappointed to a second term as dean in 2025.

## Building for Discovery

In 2021, the Abercrombie Engineering Laboratory—long the heart of engineering research at Rice—was dismantled after seventy-three years of service. In its place rose the Ralph S. O'Connor Building for Engineering and Science, a 250,000-square-foot facility supported by an extraordinary lead gift from the estate of Ralph S. O'Connor and the generosity of other philanthropic partners, with support from the Engineering Advisory Board. Opened in fall 2023, the building marked a new era for collaborative, high-impact research at Rice. Designed to accelerate discovery and innovation, the building brings together researchers from multiple departments and research focus areas. Faculty labs sit alongside shared instrumentation spaces, encouraging interactions that cut across traditional disciplinary boundaries.

Among the research hubs housed in the O'Connor Building is the Rice Advanced Materials Institute (RAMI), launched in 2023 under the direction of Lane Martin. RAMI was created to drive breakthroughs in materials science with wide-ranging applications in national security, energy, and aerospace technologies. Its mission builds on Rice's strengths in materials research and commitment to advancing

# ↗ The Deans List

## Reginald DesRoches
(2017–2020)

A civil engineer and an expert in resilient infrastructure, Reginald DesRoches joined Rice in 2017 as the William and Stephanie Sick Dean of Engineering, overseeing nine departments, 137 faculty, and 2,500 students. Under his leadership, the school grew significantly in size, research activity, and national stature. He launched key interdisciplinary initiatives in neuroengineering, synthetic biology, and data science, and established the Center for Transforming Data to Knowledge (D2K) Lab to strengthen data science education. A member of the National Academy of Engineering and the American Academy of Arts and Sciences, DesRoches later served as provost and now leads Rice University as its eighth president.

## Luay Nakhleh
(2021–present)

Since his appointment, Luay Nakhleh has led transformational growth in faculty, research activity, and academic programs. He oversaw sixty-eight faculty hires aligned with five research priorities, contributing to a 45 percent increase in research expenditures, which reached $102 million in fiscal year 2024. He worked alongside Rice leadership to expand global partnerships, and launched new degree programs in artificial intelligence, operations research, data science, energy transition and sustainability, and engineering leadership. He also led the renovation of Maxfield Hall and the opening of the Ralph S. O'Connor Building for Engineering and Science. A Sloan Fellow and Guggenheim Fellow, Nakhleh has received multiple George R. Brown teaching awards.

**Reginald DesRoches**

**Luay Nakhleh**

innovations that serve both national priorities and international challenges.

Rice also renewed its commitment to foundational academic spaces. In 2021, the Mechanical Laboratory—one of the first buildings constructed on Rice's campus—was renovated and renamed Maxfield Hall. Once home to civil and environmental engineering and mechanical engineering, the building now houses the Department of Statistics. The project was supported by a $5-million commitment from Robert and Katherine Maxfield and the Maxfield Foundation, ensuring the historic structure continues to serve future generations of students.

## Cool Water

Water has always shaped life in Houston, and Rice engineers have long led the effort to understand and manage it—whether the challenge is clean water access or excess.

In 2015, the Nanosystems Engineering Research Center for Nanotechnology-Enabled Water Treatment (NEWT) was established, thanks to over $35 million in grants from the National Science Foundation. This national center, headquartered at Rice, designs water treatment systems capable of transforming contaminated water into clean, safe drinking water. These systems are built for use in remote rural areas, humanitarian emergencies, and industrial sites. They minimize chemical and energy use, making them especially useful in off-grid and resource-limited environments. NEWT's partners include more than twenty organizations with a stake in sustainable water solutions.

That mission expanded further in 2024 with the launch of the Rice WaTER Institute. The institute takes a broad approach to water's role in public health, environmental protection, and the global transition to cleaner energy. Its goal is to address water challenges holistically, from city streets to rural villages.

"We can save more lives by having clean water than doctors can," said Pedro Alvarez, cofounder of both NEWT and the WaTER Institute.

During the early months of the COVID-19 pandemic, Rice researchers Loren Hopkins '89 '98, Lauren Stadler, and Kathy Ensor began pioneering work with the Houston Health Department to track the spread of the virus through wastewater. Their collaboration laid the scientific foundation for what would become the Houston Center of Excellence for Wastewater Epidemiology, designated by the Centers for Disease Control and Prevention in 2022 as one of the first centers in the nation. Their research demonstrated how wastewater could serve as an early warning system for community outbreaks, detecting virus levels even when clinical testing was limited. The effort united experts in public health, environmental engineering, statistics, and computational biology to develop an innovative, real-time tool for disease surveillance.

In Stadler's words, "Wastewater surveillance allows us to track infections without requiring individuals to seek testing, making it an unbiased and privacy-protecting method for public health monitoring."

Meanwhile, the SSPEED Center—short for Severe Storm Prediction, Education and Evacuation from Disasters—was focused on the problem of too much water. Led by director Philip Bedient and codirector Jim Blackburn, the SSPEED team has provided critical flood

↑ Water treatment expert Qilin Li is a professor of civil and environmental engineering and co-director of Rice's Center for Nanotechnology-Enabled Water Treatment.

→ Rice, BCarbon, and Scenic Galveston have partnered to build a living shoreline to protect coastal ecosystems in Galveston Bay.

↓ A renovated Maxfield Hall includes two new classrooms, faculty and staff offices, a conference room, and huddle rooms for collaborative projects.

forecasting and infrastructure recommendations for Houston and the Gulf Coast since 2007. Its expertise proved essential during Hurricane Harvey in 2017, when the center's hydrologic modeling and planning helped local leaders understand the unprecedented rainfall and prepare for future events.

Beyond forecasting, SSPEED is working toward long-term solutions, including developing nature-based solutions and advancing the Galveston Bay Park Plan, which proposes turning material dredged from the Houston Ship Channel into protective barrier islands that also serve as public green space.

Taken together, the work of the NEWT Center, the WaTER Institute, the Houston Center of Excellence for Wastewater Epidemiology, and the SSPEED Center tells a larger story: Rice is a leader in water. Its engineers confront the full spectrum of challenges—from global water scarcity and public health threats to coastal flooding and climate resilience—with creativity and urgency.

## Energy, Reinvented

For decades, Rice Engineering helped establish Houston's energy leadership. Today, its researchers are pioneering technologies that enable sustainable energy, address climate change, and build a more resilient global energy infrastructure.

In 2023, Rice Engineering launched the Rice Engineering Initiative for Energy Transition and Sustainability (REINVENTS) to strengthen its leadership in energy research and innovation. Led by inaugural director Aditya Mohite, William M. Rice Trustee Professor of Chemical

”

Bold ideas transformed sustainable s redefine how world uses e and materials

nto
olutions
the
ergy

### The Carbon Hub

Founded at Rice University in 2019, the Carbon Hub is transforming how the world thinks about hydrocarbons—not as fuels to be burned but as sources of clean hydrogen and advanced carbon materials. Launched with a $10-million commitment from Shell, the initiative brings together scientists, industry leaders, federal labs, and policymakers from around the world to develop scalable, sustainable solutions.

Rather than burning hydrocarbons as fuel and releasing carbon dioxide, Carbon Hub proposes splitting them to generate clean-burning hydrogen fuel and solid carbon. That carbon can then be transformed into durable materials for buildings, vehicles, textiles, and more.

Since its inception, Carbon Hub has attracted more than $12 million in additional support from the US Department of Energy, the National Science Foundation, and The Kavli Foundation. A flagship project also funded by The Kavli Foundation advances foundational nanoscience, which in turn accelerates the design of reactors that more efficiently convert hydrocarbons into solid carbon.

"We've made strong progress in building a community, advancing technology and getting the word out," said Matteo Pasquali, A.J. Hartsook Professor of Chemical and Biomolecular Engineering, professor of chemistry, materials science and nanoengineering, and director of the Carbon Hub. "We are proposing solutions that bring climate benefits while also opening opportunities to boost manufacturing and strengthen supply chain security."

→ Rice and Woodside Energy launched a five-year collaboration to reduce greenhouse gas emissions and provide lower carbon solutions. Cold plasma technology, shown here, offers a unique approach to breaking down carbon dioxide.

and Biomolecular Engineering, REINVENTS brings together engineering faculty across disciplines to drive solutions in sustainable fuels, emissions reduction, and grid resilience. It positions Rice as a strategic partner for industry, government, and global collaborators focused on decarbonization and sustainability.

That vision took a major leap forward in 2024 when Woodside Energy committed $12.5 million to create the Woodside-Rice Decarbonization Accelerator, a five-year research collaboration bringing decarbonization technology from Rice labs to market. Among the emerging solutions is a cold plasma technology that can transform greenhouse gases like carbon dioxide and methane into solid carbon, which could be used in batteries and transistors. The project is a bold step toward scalable climate solutions.

## The Power Behind Progress

From the R1 computer to today's AI-powered platforms, computing has always been at the heart of innovation at Rice. But over the past decade, its role has evolved dramatically. Computing is no longer just a tool—it's the foundation of modern engineering. It connects disciplines, accelerates discovery, and powers solutions to the world's most urgent problems.

Today, Rice Engineering and Computing is home to more than twenty faculty members advancing artificial intelligence and machine learning. Their work reflects not only technical depth but also a growing focus on ethical responsibility and societal impact. Anshumali Shrivastava, for instance, is developing energy-efficient algorithms that dramatically reduce the environmental cost of large-scale AI models. His research laid the groundwork for ThirdAI, a company that enables deep learning on standard CPUs instead of energy-hungry GPUs. Likewise, Vicky Yao develops machine learning and statistical methods that integrate data—such as single-cell studies and population-scale genome-wide association studies—to better model disease biology, enable deeper exploration, and pave the way for targeted diagnostics and therapies for conditions like Alzheimer's.

As AI plays a greater role in medicine, energy, and education, Rice researchers are working to make these systems more transparent and trustworthy. Richard Baraniuk and his collaborators developed SplineCAM, a tool that tracks neural network evolution—providing insight into what's happening inside the "black box." Vicente Ordóñez-Román tackles bias in natural language and computer vision, building algorithms that can identify and correct the amplification of stereotypes in training data. Ben Hu explores how reinforcement learning can make AI more adaptive and self-correcting when biased patterns are detected, helping safeguard fairness without sacrificing performance.

Nowhere is this more visible than in the Center for Transforming Data to Knowledge (D2K) Lab, a national model for experiential data science education supported by a generous $4-million gift from Rice alumnus Kevin Harvey and his wife, Catherine. Since 2018, the lab has paired Rice students with real-world projects from over sixty partners, from city governments to health start-ups. Whether improving cancer screening, mapping urban infrastructure, or analyzing sports performance, D2K students learn to translate complex data into actionable solutions.

Rice's momentum in computing is also driving new interdisciplinary efforts, such as building wireless infrastructure for a 6G future, developing autonomous systems for people with disabilities, and exploring quantum computing's potential. These efforts underscore a simple truth: The future of engineering is computational, and Rice is helping shape it—responsibly, creatively, and with a global view.

In the years ahead, Rice engineers will continue to expand what's possible by combining technical rigor with human insight. Whether decoding the brain, mapping global disease, predicting floods, or designing sustainable materials, computing will remain a powerful engine of innovation and a bridge between disciplines.

## Engineering Across Borders

Rice Engineering's commitment to solving for greater good doesn't stop at the edge of

campus—or the borders of the United States. Over the past decade, the school has expanded its global presence through industry immersion, academic partnerships, and collaborative research initiatives that address critical challenges in health, energy, and sustainability.

Since 2015, the Global Medical Innovation program has sent more than four hundred Rice students to Costa Rica to work alongside professionals in the medical technology industry. Focused on advancing real-world medical technologies, the program combines engineering, business, and clinical training to prepare students to design solutions for urgent global health needs.

At the institutional level, Rice Engineering has built a powerful network of international academic partnerships. Recent agreements with Tecnológico de Monterrey in Mexico, the Instituto Tecnológico de Buenos Aires in Argentina, and the University of Edinburgh in Scotland have created new pathways for dual degrees, cooperative education, and collaborative research in key areas like resilient and adaptive communities, advanced materials, and future computing.

Nowhere is Rice Engineering's global engagement more robust than in India. Beginning in 2020, the school has built a deep partnership with top Indian institutions including the Indian Institute of Technology Kanpur (IITK), the Jawaharlal Nehru Centre for Advanced Scientific Research, and the Indian Institute of Science. These efforts are advancing joint research on topics ranging from climate resilience and urban infrastructure to healthcare and data science.

Alongside institutional relationships and faculty collaborations, student academic exchange plays a central role in research at Rice. One example is the Mehta Rice Engineering Scholars Program, launched in 2023 in partnership with the Bhupat and Jyoti Mehta Family Foundation. The program brings top Indian students to Houston for hands-on research and mentorship. These students are already making contributions in wireless communication, machine learning, and drug delivery systems.

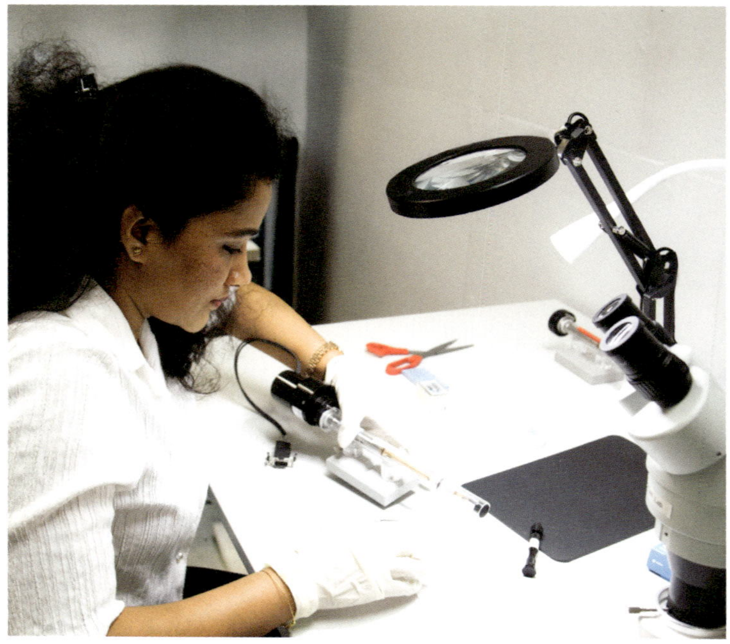

1

2

Together, these programs reflect the expanding global scope of Rice Engineering. By forming long-term partnerships and empowering students to work across borders, the school is equipping the next generation of engineers to lead on the world stage—wherever their solutions are needed most.

## Momentum into Tomorrow

In 2022, Reginald DesRoches was named president of Rice University—the first engineer to lead the institution. Two years later, under the leadership of Dean Luay Nakhleh, the George R. Brown School of Engineering officially became the George R. Brown School of Engineering and Computing. The renaming underscores the growing significance of

**1** Srilekshmi Muraleedharan, a doctoral student, conducts research in the lab.

**2** At the Data-to-Knowledge Showcase, D2K capstone teams present their end-of-semester findings, field questions, and compete for awards.

**3** In Professor Omid Veiseh's lab, doctoral student Amanda Nash—now an assistant professor at Rice—investigates innovative approaches to treating ovarian cancer.

3

computing across all engineering disciplines and acknowledges the school's leadership and excellence in the space.

This vision has been matched by momentum. Under Dean Nakhleh's guidance, the school is now home to more than 180 tenured and tenure-track faculty, over 1,600 undergraduates, and nearly 1,600 graduate students. Research expenditures have surpassed $100 million, with annual research awards approaching $150 million. The school also counts more than thirty faculty elected to the National Academies of Sciences, Engineering, and Medicine, as well as the National Academy of Inventors—an extraordinary concentration of talent and recognition. These milestones reflect more than growth. They mark a deliberate and strategic evolution in how the school educates, innovates, and serves society.

It's been quite a ride, from the first six engineering degrees awarded in 1916 to over three thousand students enrolled at the George R. Brown School of Engineering and Computing, a journey supercharged by collaboration, discovery, and the expanding power of computing.

Today, Rice engineers are at the forefront of solving global challenges—whether addressing climate risk, advancing cancer detection, tackling forever chemicals, or developing robotic limbs. What problem needs solving? The answers are right here.

At Rice Engineering and Computing, the extraordinary happens every day.

# Fifty and Forward

→ Dancing robots light up the lobby of the Ralph S. O'Connor Building for Engineering and Science during the anniversary celebration.

For more than a century, engineering has been at the heart of Rice University. On March 28–29, 2025, the George R. Brown School of Engineering and Computing marked its fiftieth anniversary, celebrating decades of groundbreaking achievements that addressed society's most critical challenges. The golden jubilee was more than a celebration. It was a call to action, inspiring the next generation to continue the tradition of solving for greater good. As the school enters its next fifty years, it remains committed to excellence in engineering education, research, and impact.

RICE

1 Engineering students showcase their latest projects at the Oshman Engineering Design Kitchen.

2 Rice President Reginald DesRoches, UC Davis Chancellor Gary May, and Dean of Engineering and Computing Luay Nakhleh launch their owls after Dr. May's keynote address.

3 A brilliant drone show lights up the Academic Quadrangle, celebrating fifty years of Rice Engineering and Computing.

1

2

**4** Charlos Ward '98 '06, Rice University Trustee and chair of the school's Engineering Advisory Board, welcomes alumni and special guests during the celebration.

**5** Professor Aditya Mohite moderates the panel Engineering Innovations in Energy and Sustainability, featuring leaders from the Greater Houston Partnership and bp.

**6** Talithia Williams '02, Rice alumna and professor of mathematics at Harvey Mudd College, highlights the importance of data fluency in the digital age.

**7** Partha Ranganathan '95, Rice alumnus and vice president and engineering fellow at Google, explores the societal implications of artificial intelligence.

**8** Rice alumnus Vivas Kumar '09, CEO and cofounder of Mitra Chem, shares his expertise during the energy and sustainability panel.

4

5

6

7

8

3

Good

## References

The historical information and quoted statements in this essay derive from the library archives at Rice University, Rice University Office of Public Affairs, and from Rice-related websites and faculty oral history interviews conducted by Emily Person, Jada Crawford, and Gabriel Witkop.

**Additional sources include:**

"AEESP Environmental Engineering & Science Stories: Herb Ward." Association of Environmental Engineering and Science Professors. Accessed December 2, 2024. https://www.youtube.com/watch?v=GL6FerdiVhw.

Alvarez, Pedro J. J. "Nanotechnology-Enabled Water Treatment: A Vision to Enable Decentralized Water Treatment and Reuse." Paper presented at NSF Nano Grantees Conference: Progress in Nanotechnology, Washington D.C., December 12, 2016.

Bedient, Philip, Alex Gordon, Ross Gordon. "Rice University Engineers Without Borders: An Exercise in International Service Learning." *American Society for Engineering Education* (2006): 11.1099.1–.14.

The Brown Foundation. "About the Brown Foundation." Accessed January 21, 2025. https://www.brownfoundation.org/about/#history.

Burrus, C. Sidney, interviewed by Frederik Nebeker. "C. Sidney Burrus, An Oral History Conducted by Frederik Nebeker." May 12, 1998. IEEE History Center, The Institute of Electrical and Electronics Engineers, Inc.

"Dr. Edwin L. Thomas." National Academy of Engineering. Accessed March 21, 2025. https://www.nae.edu/27915/Dr-Edwin-L-Thomas.

"Dr. Michael M. Carroll." National Academy of Engineering. Accessed February 8, 2025. https://www.nae.edu/29609/Dr-Michael-M-Carroll.

"Dr. Reginald DesRoches." National Academy of Engineering. Accessed April 15, 2025. https://www.nae.edu/224644/Dr-Reginald-DesRoches.

Fox, Stephen. "Planning in Houston." *Cite* (Fall 1985): 12–14.

Goodrich, Joanna. "This Rice University Professor Developed Cancer-Detection Technology: Her Optical Imaging System Eliminates the Need for a Biopsy." IEEE Spectrum. Accessed March 26, 2025. https://spectrum.ieee.org/rice-university-cancer-detection-technology.

Grossman, Max, and Vivek Sarkar. "SWAT: A Programmable, In-Memory, Distributed, High-Performance Computing Platform," 81–92. In *Proceedings of the 25th ACM International Symposium on High-Performance Parallel and Distributed Computing (HPDC '16)*. Association for Computing Machinery, 2016. https://doi.org/10.1145/2907294.2907307

Henderson, Archie. "City Planning in Houston, 1920–1930." *Houston Review* 9, no. 3 (1987): 107–36.

Hopkins, Loren., et al. "Public Health Interventions Guided by Houston's Wastewater Surveillance Program During the COVID-19 Pandemic." National Library of Medicine. July 28, 2023. https://pubmed.ncbi.nlm.nih.gov/37503606/.

"Houston Health Department Earns CDC Designation as Center of Excellence for Wastewater Epidemiology." Houston Health Department. Accessed April 14, 2025. https://www.houstonhealth.org/news/news-releases/houston-health-department-earns-cdc-designation-center-excellence-wastewater-epidemiology.

Hutchins, Shawn. "Data and Its Central Role in Pandemic Management." Houston Wastewater Epidemiology. Accessed April 14, 1015. https://www.hou-wastewater-epi.org/news/data-and-its-central-role-pandemic-management.

Iliffe, J. K., and Jane G. Jodeit. "A Dynamic Storage Allocation Scheme." *The Computer Journal* 5, no. 3 (November 1962): 200–209.

Kean, Melissa. "Art Busch and The Birth of Environmental Studies at Rice." *Rice History Corner* (blog). Accessed December 2, 2024. https://ricehistorycorner.com/2021/03/24/to-anticipate-and-plan-for-problems-of-the-future-art-busch-and-the-birth-of-environmental-studies-at-rice/.

Kean, Melissa. "Bonner Lab." *Rice History Corner* (blog). Accessed February 24, 2025. https://ricehistorycorner.com/2011/08/11/1694/.

Korschun, Holly. "Rice's McIntire Named Chair of Bioengineering for Emory, GT." Emory Report 55, no. 28 (April 21, 2003).

Littlefield, David. "Abercrombie, James Smither." Handbook of Texas Online. Accessed November 1, 2024. https://www.tshaonline.org/handbook/entries/abercrombie-james-smither.

Lovett, Edgar Odell. "Houston a Hospitable Home for the Builders of Rice Institute." *The Houston Post*, September 21, 1922, page 14.

Martin, Douglas. "W. E. Gordon, Creator of Link to Deep Space, Dies at 92." *The New York Times*, February 28, 2010, section A, page 24.

McGrath, Dylan. "Gene Frantz: Digital Signal Processing Visionary." *Electronic Engineering Times*. Accessed November 18, 2024. https://www.eetimes.com/Gene-Frantz--Digital-signal-processing-visionary/.

"NASA Specialized Center of Research and Training (NSCORT) in Gravitational Biology." Rice University Institute of Biosciences and Bioengineering. Accessed March 14, 2025. https://ntrs.nasa.gov/api/citations/19970009628/downloads/19970009628.pdf.

"Mary F. Wheeler." Oden Institute for Computational Engineering and Sciences, University of Texas at Austin. Accessed February 15, 2025. https://oden.utexas.edu/people/directory/mary--f.--wheeler/.

McIntire, Larry V. "J. David Hellums." In *National Academy of Engineering, Memorial Tributes: Volume 24*. The National Academies Press, 2022. Accessed January 5, 2025. https://doi.org/10.17226/26492.

"Michael M. Carroll." Dignity Memorial. Accessed February 8, 2025. https://www.dignitymemorial.com/obituaries/bellaire-tx/michael-carroll-6768325.
"Remembering Ken Kennedy."

*SCIDAC Review, U.S. Department of Energy Scientific Discovery Through Advanced Computing* (Fall 2007): 51–61.

"Rice University Announces New Leadership and Plans to Again Expand Enrollment." *Inside Higher Ed*. Accessed April 15, 2025. https://narratives.insidehighered.com/rice-university-new-leadership/index.html.

Sollie, Moriah. "Houston Works Toward Data-Driven Disaster Response." *Government Technology*. Accessed April 15, 2025. https://www.govtech.com/public-safety/houston-works-toward-data-driven-disaster-response.html.

**Additional Photo Credits:**

Atanu Purkayastha: pp. 56-57
Brandon Martin: pp. 40, 120, 123, 128
Frank Brevoort: p. 55
Frank J. Schleuter: p. 36
Frank van Orden: p. 25
Jorge Vidal: p. 128
Luxe Studio Productions: pp. 138-140

## Acknowledgments

This book would not have been possible without the shared enthusiasm of a remarkable community—faculty, students, staff, and alumni—who brought their time, insights, and memories to the table. What began as an effort to capture history quickly grew into a broader celebration of culture, collaboration, and impact. We are grateful to all who helped shape this story—on the page and behind the scenes.

Oral interviews were conducted in the Digital Media Commons (DMC) at Rice University's Fondren Library, with generous support from the DMC staff. These conversations—led by Emily Person of the School of Engineering and Computing's marketing and communications team—will be archived to help preserve the school's living history.

**We extend our heartfelt thanks to the following individuals who shared their reflections through interviews:**

Keith Cooper, L. John and Ann H. Doerr Professor in Computational Engineering; Professor of Computer Science

Don Johnson, J.S. Abercrombie Professor Emeritus, Electrical and Computer Engineering; Professor Emeritus, Statistics

Ed Segner, Professor in the Practice of Civil Engineering Management

Laura Schaefer, Burton J. and Ann M. McMurtry Chair in Engineering; Professor, Mechanical Engineering

Pedro J. J. Alvarez, George R. Brown Professor of Civil and Environmental Engineering; Director, NEWT Center; Director, Rice WaTER Institute

Philip Bedient, Herman Brown Professor of Engineering; Director, SSPEED Center

Gene Frantz, Professor in the Practice, Electrical and Computer Engineering

Matteo Pasquali, A.J. Hartsook Professor of Chemical and Biomolecular Engineering; Professor of Chemistry, Materials Science and Nanoengineering; Director of the Carbon Hub

Antonios Mikos, Louis Calder Professor of Bioengineering and Chemical and Biomolecular Engineering; Professor of Chemistry; Professor of Materials Science and Nanoengineering; Director of the Biomaterials Lab, Center for Excellence in Tissue Engineering, and J.W. Cox Laboratory for Biomedical Engineering

Kyriacos Zygourakis, A.J. Hartsook Emeritus Professor, Chemical and Biomolecular Engineering

Bart Sinclair, Emeritus Professor of Electrical and Computer Engineering

Kathy Ensor, Noah G. Harding Professor of Statistics; Director of the Center for Computational Finance and Economic Systems

Richard Tapia, University Professor; Maxfield-Oshman Chair in Engineering; Professor of Computational Applied Mathematics and Operations Research; Faculty Director of the Tapia Center

Mary Wheeler, Professor Emeritus, Aerospace Engineering and Engineering Mechanics, Petroleum and Geosystems Engineering, The Oden Institute for Computational Engineering and Sciences at The University of Texas at Austin

We are grateful to Portia Hopkins, current Rice University Historian, for her support and insights during the development of this project.

**Special Thanks**

A special thank-you goes to Melissa Kean, who served as university historian for more than thirty years. Melissa shared not only a wealth of research and resources but also firsthand accounts that helped bring key moments in Rice's engineering story to life.

We also appreciate the contributions of the Fondren Fellows Program, which supported early research for this project. Thank you to engineering students Patrick Bednarz, Gabriel Witkop, and Lexi Xu, and their mentor Amanda Focke of Fondren Library, for laying the foundation for what this book would become. We are grateful to the staff of the Woodson Research Center for their invaluable guidance and hands-on support.

The School of Engineering and Computing's marketing and communications team—Jada Crawford, Maria Aglietti, Emily Person, and Veronica Tremblay —guided this project from start to finish, shaping its tone and structure.

We are also thankful for the partnership of Cynthia Riley and Katie Donovan, who supported this project with thoughtful feedback and attention to detail.

And finally, we are deeply grateful to author Joe Nick Patoski; to DJ Stout and Connor O'Neill of Pentagram Design for shaping the book's visual identity; and to Heather Brand, whose editorial perspective brought clarity and cohesion to each chapter.

This book is the product of many minds working together—and it is richer for it.

Published in 2025 by Rice University
School of Engineering and Computing

Distributed by
Texas A&M University Press

Edited by
Heather Brand,
Jada Crawford, Emily Person,
and Veronica Tremblay

Design by Pentagram
DJ Stout and Connor O'Neill

Printed in Dallas, Texas
Blanks Printing and Imaging, Inc.

Unless otherwise indicated, all photographs are courtesy of Woodson Research Center, Fondren Library, Rice University, Jeff Fitlow and Gustavo Raskosky of the Rice University Office of Public Affairs, and Donald Soward of Rice University School of Engineering and Computing.

ISBN
979-8-218-76421-0

Library of Congress Control Number
2025920284